Dangerous Discourses of Disability, Subjectivity and
Sexuality

Dangerous Discourses of Disability, Subjectivity and Sexuality

Margrit Shildrick

Queen's University Belfast, UK

First published 2009 by
PALGRAVE MACMILLAN

Palgrave Macmillan in the UK is an imprint of Macmillan Publishers Limited,
registered in England, company number 785998, of Houndmills, Basingstoke,
Hampshire RG21 6XS.

Palgrave Macmillan in the US is a division of St Martin's Press LLC,
175 Fifth Avenue, New York, NY 10010.

Palgrave Macmillan is the global academic imprint of the above companies
and has companies and representatives throughout the world.

Palgrave® and Macmillan® are registered trademarks in the United States,
the United Kingdom, Europe and other countries.

ISBN-13: 978-0-230-21056-1 hardback

This book is printed on paper suitable for recycling and made from fully
managed and sustained forest sources. Logging, pulping and manufacturing
processes are expected to conform to the environmental regulations of the
country of origin.

A catalogue record for this book is available from the British Library.

Library of Congress Cataloging-in-Publication Data
Shildrick, Margrit.
  Dangerous discourses of disability, subjectivity and sexuality /
  Margrit Shildrick.
    p. cm.
  Includes bibliographical references and index.
  ISBN 978-0-230-21056-1 (alk. paper)
  1. People with disabilities. 2. Marginality, Social. I. Title.
  HV1568.S454 2009
  305.9'08—dc22   2009013616

10   9   8   7   6   5   4   3   2   1
18   17   16   15   14   13   12   11   10   09

Printed and bound in Great Britain by
CPI Antony Rowe, Chippenham and Eastbourne

Contents

Acknowledgements

The manuscript is complete and ready to send off to the publisher, my study floor is finally cleared of accumulated reference material, and now comes the always difficult task of writing the acknowledgements. I tend to write in extended, intense, and isolated stretches, but so much intellectual, emotional, and even physical work goes into getting to this stage of the process that nothing I say here will be fully adequate.

During the years I have been pursuing this project I have been fortunate enough to have travelled and worked or held research positions widely beyond my home location, and I'm grateful to all those who provided time, space, and funding. My thanks go especially to Ailbhe Smyth of University College Dublin for her unwavering support, and to all the participants in the three years of the open seminar series, Advanced Feminist Theory, which provided the opportunity for me to try out some of the nascent ideas that later developed into chapters in *Dangerous Discourses*. During much of the same period I was co-editing a book on feminist corporeal ethics with Roxanne Mykitiuk of Osgoode Hall Law School, Toronto, who made the inspired suggestion that I take up a visiting research scholarship at Cornell Law School under the direction of Martha Fineman, who is well-known as a great supporter of disability scholarship. The Toronto connection itself produced further openings, including a wonderful semester teaching at York University on the newly convened Masters in Critical Disability Studies. The director of the programme, Marcia Rioux, despite her own very different approach to disability, gave me all encouragement to work with the students in a way that surely overturned some cherished doxa. The vigorous response of the course participants challenged me to think more clearly and effectively, to the extent that it remains my own most valuable period of teaching. Since then, as an adjunct professor of disability studies at York, I have supervised several excellent graduate dissertations with outstanding contributions from Loree Erickson, Julie Devaney, Jen Paterson, and Kelly Fritsch. A series of visiting fellowships at HCTP and Centre for Women's Health – both institutions of University of Toronto – brought me into close contact with Pat McKeever and Gillian Einstein, both highly innovative thinkers; while time at the Centre for Somatechnics at Macquarie University and at Sydney Law School was enabled by Nikki

Sullivan and Isabel Karpin respectively. All those mentioned, and many more, have played a significant role in the process of my writing this book.

Back in the United Kingdom, my major thanks on both an intellectual and personal level are due to Janet Price who has collaborated with me on numerous presentations and publications. Her willingness to pursue all sorts of unorthodox ideas has been a mainstay in the development of the modes of thinking that inform *Dangerous Discourses*, though I hesitate to say she would agree with it all. Many other friends and colleagues have contributed in multiple ways, including Holliday Tyson and Nancy Halifax who have made Toronto home from home. Finally, the project simply would not have come to fruition without the ongoing and unquestioning support and generosity of my partner, Lis Davidson.

As with most single-authored books, some of the material has already seen the light of day in other forms. Several chapters are based on previously published articles, which have been to a greater or lesser extent reworked and recontextualised, and often distributed over two or more locations. In the era of postmodernity and cyberspace nothing is ever final, and I have incorporated some work from my own prior publications in *Cultural Studies* 2005, 19. 6, *Critical Quarterly* 2005, 47. 3, *Sexuality Research and Social Policy* 2007, 3. 3, *SCAN* 2004, 1.3, *Studies in Gender and Sexuality* 2007, 8. 3, and from a joint paper with Janet Price that appeared in *Rhizomes* 2005, 11. I gratefully acknowledge all permissions granted.

Introduction

Why is it that at the beginning of the twenty-first century – with its multiple geo-political insecurities and anxieties, its distinctly ambivalent expectations of the future, and its growing awareness of internal pressures – the western world and its developed counterparts should be so unsettled by anomalous embodiment? What is it about the variant morphology of intra-human difference that is so disturbing as to invoke in the self-defined mainstream not simply a reluctance to enter into full relationship, but a positive turning away and silencing of the unaccepted other? For such a disengagement to persist in the face of apparently more weighty global concerns speaks not to an over-investment in the local and individual, a kind of displaced anxiety even – though that cannot be entirely discounted – but to the extraordinary significance of human corporeality. To be named as differently embodied is already to occupy a place that is defined as exceptional to some putative norm, rather than to simply represent one position among a multiplicity of possibilities. The self-evidence of differential form across human morphology as a whole is put aside in favour of a discourse in which some people are taken to inherently exceed the boundaries of what counts as normative embodiment. In this text, I concentrate on the continuing discursive exclusion of disability within western and western-inflected societies, and argue that at the very same time that such states are making tremendous strides towards the *formal* integration of disabled people into the rights, obligations, and expectations of normative citizenship, a counter-trend of segregation is equally in play. The title of this book – *Dangerous Discourses* – is no rhetorical flourish, but an indication of the depth of anxiety that engagement with disability elicits.

My thesis is that disabled people[1] continue to endure broad cultural discrimination and alienation, not so much for their difference (which

1

may of course be hidden) but because their form of living in the body lays bare the psycho-social imaginary that sustains modernist under-standings of what it is to be a subject. Where physical and mental autonomy, the ability to think rationally and impartially, and interper-sonal separation and distinction are the valued attributes of western subjectivity, then any compromise of control over one's own body, any indication of interdependency and connectivity, or of corporeal instability, are the occasion – for the normative majority – of a deep-seated anxiety that devalues difference. In an earlier book, *Embodying the Monster*, I traced that negative trajectory through the history of the monstrous into present-day manifestations that implicate a variety of non-normative forms of embodiment: from conjoined twins, through displaced persons like refugees, to the futuristic cyborg. Yet for all the narrative of violence, oppression, and disavowal of the monstrous, the coincident exposure of a motivating and irreducible vulnerability at the heart of all human embodiment signalled not the inevitability of rejection, but the possibility of what Haraway (1992) terms a 'regenera-tive politics'. Her opening up of 'the promises of monsters' gives hope to a different future. In taking up my present focus on states of disability, I reiterate some of the same themes, but in a more substantive and every-day context that imbricates with, yet fundamentally critiques, a recent history of disability politics that has also sought to overturn the norma-tive paradigms that determine who shall be valued and who not. Against the standard demands for an extension and solidification of rights for disabled people, and for a more thoroughgoing inclusion of their spe-cific interests in cultural production, I turn instead to an investigation of what it is that continues to impede the evolution of equitable conditions of possibility. Such a mode of thinking marks what is often termed 'crit-ical disability studies', a relatively recent development – within which I situate my own work – that is broadly aligned with a postconventional theoretical approach. Its purpose is both to extend into new territory the achievements of working through a more modernist paradigm of disability, and where necessary to productively critique the limitations of that model.

Before proceeding further it is necessary to ask what comprises the category of disability, or what I have less contentiously referred to as anomalous embodiment. Although some form of definitive answer is often called for – in academia no less than in terms of a general public – it is one that I, along with many other critical disability theorists, and particularly those working within a postconventional framework, am reluctant to provide. To set out any mutually agreed series of parameters

University of Plymouth Library

Subject to status this item may be renewed
via your Voyager account

http://voyager.plymouth.ac.uk
Tel: (01752) 232323

would be to close down, and thus normalise, what must otherwise remain a shifting nexus of both physical and mental states that resists full and final definition. On the simplest level, what counts as disabling anomaly varies greatly according to the socio-historical context,[2] and even were the enquiry limited to a westernised location in our own time, the category remains multi-faceted and hard to pin down. As Judy Rohrer notes:

> There are differences in type of disability (in a reification of the mind/body split, disability is usually broken down as physical or intellectual), in impact (minor hearing loss versus paralysis), in onset (disability from birth/gradually becoming disabled/suddenly becoming disabled), in perceptibility (having a 'hidden disability' and 'passing' as non-disabled versus being unable to hide a disability), in variability (most disabilities change across time and space), and in prevalence (disabilities vary by sex, ethnicity, age, and environment).
> (2005: 41)

None of that variation prevents the kind of universalising approach that continues to speak of disability as a single classification, and to a certain extent that move is both inevitable and appropriate. In making the point that those with disabilities are 'othered' within past and present western societies, reference is clearly made to the binary logos that underlies the structuration of all such societies in the modern era. For just that reason, the contestation of ableist attitudes, values, and politics will often set aside intricate differential considerations in the face of strategic necessity. As with any socio-political challenge, at least in its initial stages, strength is often vested in presenting a unified group definition and having a common cause, even though that may mean covering over internal differences. And in a similar way, to speak of the problematic of disability in theoretical terms must both respond to, as well as take apart, the dominance of binary thinking. Regardless of how far modernist binaries may be deconstructed, regardless of how unstable their bare understanding of oppositional difference, they cannot be simply dismissed: their power may be based on an illusion, but its operation is all too real. What matters is that we do not mistake the challenge to the *effects* of binary opposition as the limit of what is possible and necessary.

There are two apparently distinct but ultimately related issues that should be stressed here. The first is that while the insistence on, and sometimes strategic suspension of, the multiple subdivisions of disability may enrich our understanding by acknowledging that each requires a different description, the further point is that the parameters around all and any types of disabling condition are in any case uncertain. In the postconventional approach, all categories are slippery, fluid, heterogeneous, deeply intersectional, and thus resistant to definition. At the same time, and this is the second crucial point, in recognising that what constitutes the 'otherness' of disability is hard to pin down in any substantive way, we should not allow the more complex sense in which the binary distinction between disabled and non-disabled is itself contestable along deconstructive lines to be obscured. Lennard Davis, for example, encapsulates one highly significant aspect of the issue when he speaks of 'the instability of the category of disability as a subset of the instability of identity in a postmodern era' (2002: 25). Indeed, as I have already indicated, one of the major themes of this book will centre on the question of why disability should be so unsettling, an effect that surely speaks not to an absolute difference between those who are and those who are not disabled, but rather to a deeply disconcerting intimation of commonality. Yet nothing in the structure of western culture – whether that is taken to refer to material or discursive operations – prepares us for such an insight which has the capacity to deeply disturb all questions of self-identity. Henri-Jacques Stiker puts it at its most provocative and personal when he comments: 'Each of us has a disabled other who cannot be acknowledged' (1999: 8). None of this means that the term 'disability' – and its various subcategories – is beyond use, but only that it must remain always open to question, indeed always pose the question of what would constitute 'non-disabled' (Price and Shildrick 1998). Although many disability advocates use the term 'temporarily able-bodied' (TAB) to describe the normatively embodied and to remind them of the impermanence of their morphological status, the convergence of categories is more profound. As Thomas Couser puts it: 'Part of what makes disability so threatening to the non-disabled then may be precisely the indistinctness and permeability of its boundaries' (1997: 178).[3]

As Davis and Couser indicate, it is the difference of disability that is at stake, but it is not sufficient to put in place formal structures of equality to ameliorate the situation. There are of course many pragmatic aspects of living with a disability that can benefit from such an approach, but they have limited application in terms of both the embodiment of difference and the unease that disability so often generates. Accordingly, as will become clearer in the course of my extended argument in the chapters following, I have grave reservations about the social constructionist model of disability (SMD), and prefer instead to enquire into both the phenomenological experience of the disabled body and the psycho-social dimensions of what constitutes an excluding normativity. Where I follow the SMD lead, however, is in locating the major 'problem' with the normative mainstream. Over the last couple of decades, the disability movement in both North America and the United Kingdom has undergone a tremendous change in realigning the direction of its endeavours from challenging the biomedically driven discourse of disability as an inherent individual failing of physical or mental development to understanding that the condition – as a universal category at least – is a social construction. The clear implication is that were social and political structures organised differently to take account of embodied anomaly, then disability itself would lose its negative status, or indeed, in the view of some theorists, be reduced to a neutral impairment of little intrinsic significance. The prevalence of the SMD, in the United Kingdom, in particular, where it was initially developed, has resulted in some considerable material gains for disabled people insofar as many western states have instantiated new legislation to undercut discrimination. That in turn has undoubtedly promulgated a more inclusive organisation of social life, without, however, necessarily contesting the underlying attitudes, values, and subconscious prejudices and misconceptions that figure an enduring, albeit often unspoken, intolerance. That intolerance can be found in non-disabled and disabled people alike, not because the latter have turned against a self-identity that aligns them with an oppressed minority, but because both categories share a psycho-social imaginary that disavows morphological imperfection. The issue, then, is not simply to go on extending the formal framework in which people who experience disabilities can maximise their own potentials, but to seek ways of first understanding and then transforming the nature of that imaginary. In emphasising the problematics of subjectivity and sexuality, or more succinctly the question of the disabled sexual subject, I aim to interrogate related areas in which the normative is most firmly entrenched and therefore most challengingly deconstructed. It is a matter of both discursive analysis and substantive intervention for, as I go on to argue, disability quite fundamentally performs a queering of normative paradigms.

What, then, constitutes a dangerous discourse is any coming together of disability with either or both subjectivity or sexuality, where the perceived danger may lie equally in two related directions. In the first place there is a disruption of the perceived stability of normative expectations, and in the second a testing of the doxa that has directed disability politics primarily to the reform of an evidently harmful external social structure. In both instances, what is at stake is an entrenched view that any understanding of the operation of lives across differences entails an appreciation of distinct modes of being, which in turn relies on a clear sense of where the definitional boundaries lie. By contrast, my own project intends to unsettle the conventions on both sides of the putative divide between disabled and non-disabled, and to look afresh at how and why definitions are put in place. This is no easy task as theorists and activists alike – however they identify personally – are often deeply invested in a characteristically western conception of the world as grounded in binary opposites that seem to speak unproblematically to a socio-politics of having or not-having, and to inclusion or exclusion. The problem, as I understand it, is that at some deep level the very act of identifying with one category rather than the other is seen as a source of strength and stability that requires no further analysis. For those who are able to align themselves with normative forms of embodiment, the external goods, benefits, and advantages that come their way are simply entitlements, which they may or may not wish to extend to others, while for those named as disabled the issue is to gain access to what they are denied. Where jobs, education, or housing, for example, are concerned, the struggle for equal opportunity may be painful and prolonged, but it is one in which the antagonists 'know their place' and can speak from it without the need for much *self*-reflection. The losses and gains may cause a degree of disruption to either side that can reformulate categorical assumptions, but the fundamental binary of disabled/non-disabled is undisturbed. The issues of both subjectivity and sexuality, however, go right to the heart of what it is to be a self at all. They are productive of anxiety precisely because they displace normative and shared assumptions about the correspondence between bodily markers and the status of the self. And insofar as they generate demands, not so much for reform, as for a transformation in the meaning of selfhood for every one of us, they threaten the very basis of a comfortably taken-for-granted personal and social existence. In short, the interweaving of disability, subjectivity, and sexuality constitutes a dangerous mix. And it is all the more so for being both discursive and substantive.[4]

The meanings both of subjectivity and sexuality are by no means self-evident, and I shall discuss their varying and sometimes contradictory uses in the course of the chapters following. For now, it suffices to say that even as the SMD posits a social constructionist understanding of disability, it assumes nevertheless that there is some underlying pregiven subject impervious to change, and waiting to be empowered. It is hardly surprising, then, that where disabled people have been so often treated as passive objects of concern, rather than as subjects in their own right, the question of self-directed agency should become a political issue. Yet just as feminism has had to come to terms with the deconstruction of the concept of subjectivity, even as the pragmatic push has been to claim full subject status for women, so too disability studies needs to ask whether buying into the existing system – as though the problem were simply one of exclusion – is the best way forward. If the subject is no longer the stable and grounding category that it was once assumed to be, if it too is a discursive construction, then all sorts of epistemic and ontological claims must be rethought. In a similar way, the notion of sexuality is equally disputed, both with regard to what it is actually supposed to denote – a set of practices, an identity, a relationship to gender – and in terms of the claims that might be made against it. Given that both cultural and queer theory are deeply invested in problematising what is meant by sexuality, it is surely incumbent on disability studies to complicate the socio-political assertion that disabled people have the same rights as everyone else to sexual expression. In any case, as I shall discuss, the issue devolves not only on the restrictions that are placed on the anomalous body, but on the anxiety-provoking coming together of disabled and non-disabled sexual desire and practice. And if sexuality is already a dangerous discourse that threatens always to disrupt the smooth and predictable organisation of social relations – a threat that is heightened in the context of disability – then we need to ask too what psycho-social factors are in play and what exactly is the status of the boundaries that are in danger of being transgressed.

Without doubt, the most influential and taken-for-granted boundary of all is the one that separates those who count as ablebodied from those who are marked as disabled, and yet, like the definitions of subjectivity and sexuality, neither of those terms is as self-evident as they appear to be. Where in *Embodying the Monster* (Shildrick 2002), I laid out the general conditions in which the apparently external category of the monstrous other is fully imbricated with the self, the specific focus in this project makes even clearer that the separation and distinction between diverse morphologies is at best a convenience, and at worst a violent imposition, both literally and metaphorically, of epistemic power. Despite a long-term flirtation with identity politics that seems to offer the most socio-political leverage, to disability activism at least, the organising binary of ablebodied and disabled is one that has been increasingly challenged in recent years within the field of progressive disability studies. In rejecting the conservative agenda that disabled people are a distinct group who nevertheless are entitled to all the rights and benefits of their particular society, or at least to compensation where those cannot be accessed, many theorists now subscribe both to the notion of difference, and to the blurring of boundaries at the edges, in such a way that problematises the whole categorical distinction. The point that is most commonly made is that over a lifetime any individual may experience unexpected trauma, incapacitating disease, or simply the processes of ageing that can result in disability. In other words, the status of ablebodied is at best provisional rather than a fixed identity. There is nothing exceptional in such a claim that any one of us may cross the boundary between one category and another, but nonetheless I find it deeply inadequate to my own project. My interest lies in uncovering the imbrication *within* difference that weaves together two apparently distinct forms of embodiment. The problem is not only that the traffic in the mainstream model is all one way, thus implying a falling away from what remains a normative standard, but that it says nothing of either material or psychic intercorporeality.[5] To the extent that I agree that the 'able' body is always in transition to something less declarative of fullness and wholeness, I use the term 'non-disabled' to mark my own understanding of the binary difference that nevertheless pervades normative structures of thought. And where the grounds for disconnection and categorical demarcation are more directly contested, I attempt to capture that uncertainty in the expression 'dis/abled'. None of these is entirely satisfactory, but they do indicate the necessity for a certain fluidity that breaks with past paradigms.

My argument, moreover, is that every one of us, regardless of our own personal form of embodiment, is implicated in the nature of the sociocultural imaginary that shapes our everyday lives, and that as such the challenge to, and of, the discourse of disability is a shared responsibility. Given its relatively recent emergence as a real force for change, disability politics, and latterly disability scholarship, has somewhat lagged behind the sophistication that is now exhibited in either feminist or postcolonial discourse. For a long time, the received orthodoxy has been that only those people who are themselves disabled in some way have a right to speak authoritatively in the area, or to make demands that impinge on the way in which people with disabilities experience their own lives. The wholly understandable and historically verifiable claim is that for far too long the ablebodied majority has taken on, or rather usurped, the task of speaking about disability to the extent that those who experience it first-hand have been silenced and invalidated. As with early second-wave feminism, and indeed other liberatory theories, the attraction of what is called standpoint theory is that it openly privileges the lived experience and knowledge of those at the centre of the problematic, and therefore gives a voice to those who may previously have been unheard. The intention of standpoint theory is not, however, to multiply the epistemic fields available to analysis, but to promote as it were a hierarchy of truth-telling in which the putatively powerful – effectively the historic oppressors – are not to be trusted. Following Foucault, there is, I think, no difficulty in accepting that power constructs a very partial form of knowledge, but as Foucault (1980) would also assert, the distortion of truth does not imply some absolute standard there to be excavated. In other words, we should be deeply suspicious of any claim that disabled people speak the truth of their own conditions. In contemporary feminist thought, standpoint theory has all but disappeared as a viable mode of operation, but it appears to maintain a hold in disability theory and practice. There is clearly good reason to promote previously discounted voices, but not to the extent of claiming a unique authority that denies the veracity of any account or theory proposed by those who are defined as ablebodied. On the contrary, I claim that those very people have a precise responsibility in the matter, not to speak on behalf of others unlike themselves, but to interrogate their own cultural and psycho-social location as non-disabled.

It will be clear by now that my own position is marked by the desire to bring a variety of theoretical resources that are not commonly deployed in mainstream disability studies to bear on the problematic. Although it would be a mistake to simply conflate the arena of enquiry with other discourses of difference, like those pertaining to race or gender, there are sufficient points of commonality to warrant some cross-cutting strategies. In any case although disability itself encompasses enormous variations – from the very evident differences between physical and developmental anomalies to the specific and precise degree to which the 'same' condition will operate in inconsistent ways in two individuals – there is a real need for a comprehensive mode of understanding the status and significance of the overarching terms that we all deploy. The ideal might be to interrogate the particularity of every given instance of disability, but what I offer provides, I hope, at very least a starting point. Although the links often remain implicit, my own approach has been most influenced by postmodernist feminist thought – another dangerous discourse that has left little unchanged – but it takes up, too, many other perspectives that do the work of critique. Accordingly, I deliberatively utilise – and sometimes unashamedly deform – elements of variously feminist, queer, poststructuralist, and postmodernist theory to effect a transdisciplinary disruption of the conventional meanings of the terms associated with disability, subjectivity, and sexuality. What runs through all those discourses is a retheorisation of the question of difference that itself relies on thinking through the implications of the shift from the modernist understanding of the embodied self as autonomous and stable to the postmodernist insistence that the self is always interdependent, fluid, and endlessly in process. What is required then is a thoroughgoing critique of existing conventions that rather than instituting some form of successor theory is content to explore temporary theoretical conjunctions that hold open the multiplicity of potentially contradictory meanings and significations. The aim is not to discover some universal theory of disability that will ameliorate the conditions in which it is lived, but to position it as a fundamental and irreducible challenge – on diverse levels – to the normative desire to establish a certain security and predictability about the nature of human being. Indeed, the question that is implicitly posed throughout the following chapters is: what it would mean, ontologically and ethically, to reposition dis/ability as the common underpinning of all human becoming?

The grounding chapter of *Dangerous Discourses* takes up the challenge, then, at the level of corporeality to explore how and why the disabled *body* disrupts the hegemony of normative socio-cultural determinations by necessarily falling outside the conventions of bodyliness in western modernity. Where modernist stereotypes of disability construct and perpetuate a damaging and specific devaluation of some aspects both of bodily difference and of disabled people themselves, contemporary theories of embodiment uncover the uncertain nature of all embodied selfhood, including affect, emotion, and, not least, sexuality. In contrast to the constitution of a supposedly secure subjectivity through the exclusionary strategies of bodily abjection, separation, and distinction, both the insights of phenomenology and the implications of the performativity are explored to suggest how the hierarchy that overvalues some forms of embodiment at the expense of others might be productively disrupted. The implications of morphological difference are then contextualised in the next chapter within the transhistorical themes and contradictions that underpin the othering of people who experience disability. With the limited parameters of the socio-political demands of disability activism in mind, I turn to a perspective inspired initially by Foucault (2003a) and Henri-Jacques Stiker (1999), to show how the received history of the disabled body is deeply problematised by a genealogy that disrupts the notion of a continuous development of ideas and images, and shifts the focus to competing, fractured, and discontinuous discourses culturally embedded in particular historical periods. But the complication of the linear transformations of meaning charted by mainstream disability studies – from early 'religious' models concerned with a god-given nature to a biomedical approach that pathologises the disabled body, and more recently to the social constructionist analysis that effects a politicisation of the problematic – does not go far enough. In addition to the genealogical approach, it is, I argue, necessary to consider a quite different current of interior and less accessible responses that not only challenge the privileging of periodisation in historical thought, but equally disorder the alternative Foucauldian model. Where his analysis is concerned with the multiple manifestations of the self/other binary, a more specifically deconstructive approach uncovers the other as an interior attribute of the embodied self. The chapter brings some historical moments together with postmodernist theory to show how binary differences are constantly undone by the irreducible *différance* of the disabled body.

Chapter 3 moves to the heart of the project to engage with the complex issue of who may count as a sexual subject. Using a postmodernist analysis I explore the largely untheorised conjunction of disability and sexuality, and address the disqualification of non-normative bodies from discourses of sexual pleasure and desire. Given that the inter-corporeality of the sexual relation is in any case a potential point of disturbance to western normativities, the manifestation of corporeal difference in the context of sexuality appears especially anxiety provoking and threatening. In consequence, there is an extraordinary reluctance to acknowledge that disabled people have any sexuality at all, with the result that their sexual expression is highly regulated, if not invalidated or silenced completely. In exploring the cultural, political, and legal restrictions on the desires of those positioned outside the norms, I identify a thoroughgoing governmentality at the heart of institutional responses that ensures that policy initiatives are never as straightforwardly progressive as they may seem. All rights-based equality claims arising from a recognition of the limits of the modernist model of sovereign selfhood, and more specifically from the suppression of sexual subjectivity for disabled people, are double-edged in their putative appeal, offering material gains at the cost of normalisation and the covering over of difference. Moreover, I argue that the whole question of sexual citizenship, which has recently been strongly promoted within disability politics, is fraught with dangers and can make no move towards radically contesting and transforming the understanding of sexuality in contemporary society. In short, the appeal to the normative order – in whatever form it might take – cannot break the devaluation of difference. More fundamentally, although the chapter focuses on the operation of external discursive power, and traces the challenges to it that might act as a limited motor of change, it is never forgotten that such perspectives cannot take account of the psychic significances of irreducible differences in embodiment.

Consequently, my appeal – developed in 'Subjectivity, sexuality and anxiety' and subsequent chapters – for a more nuanced understanding of the emergence of normative constructions of sexuality supplements the preceding analysis by reconfiguring the problematic to engage with the psychic elements at play in the western imaginary. Building on previous work on the meaning of the monstrous (Shildrick 2002), I explore the notion of anxiety within the specific parameters of the pleasure and danger of sexuality and erotic desire, and develop a psychoanalytic approach to ask what part the link between desire and lack plays in thwarting a positive model of disability and sexuality. Despite the widespread, and often justified, wariness of theorists of disability towards psychoanalysis, I see it as offering an important way of understanding that our apparent psychic and bodily integrity is a process that is at best provisional and open to disruption, and thus always risks the breakthrough of anxiety, especially with regard to sexuality. In particular I turn to the psycho-cultural significance of the inherent dis-integration of the Lacanian infant body, and to the subsequent emergence in the Symbolic of a coherent sexual subject, shaped by the Law of the Father. What has been repressed in order to achieve unity and order, and which forms of embodied subjectivity cannot be countenanced? The approach offers some intriguing and hitherto underexplored insights into the socio-cultural denial of desire in the context of the encounter between normativity and disability, but I conclude that the Lacanian model cannot escape a certain impasse with regard to disabled people themselves, and that other non-psychoanalytic models may prove more productive.

The next chapter leaves behind the specifics of sexuality and the apparent stasis represented by the Law of the Father, to move on to the territory of juridical law. Drawing mainly on the work of Michel Foucault (2003a) and Jacques Derrida (1995, 2000), the chapter explores the critical intersection between their highly individual readings of the discursive significance of law and asks how it might relate to the substantive ground of transgressive bodies. Although neither is directly concerned with disability, I extend their speculations on the meaning of the monstrous in general to take account of the problematic parameters of the disabled body. In his text *Abnormal*, Foucault (2003a) claims that the monstrous is constituted not by the otherness of morphological abnormality as such, but by the offence it offers to the law in transgressing the established customs and regulations that set out what is proper. This juridical framing enables Foucault to identify conjoined twins as the privileged signifier of the early modern meaning of the monstrous, precisely because the figure of two in one is uncontainable within the normative structures of civil, canon, or religious law. But where Foucault makes a distinction between monstrosity and disability on the grounds that the latter – unlike the former – may be encompassed and contained within legal frameworks, contemporary disability studies increasingly deploys queer theory to reopen the question of transgression across all designations of anomalous embodiment. It suggests – contrary to Foucault – that everyday law can never resolve the problematic of disability, and that what is at stake is no longer the place of normativity and governmentality, but the sphere of an impossible justice. In the terms used by Derrida, the imbrication of the 'monstrous *arrivant*' and the Law – as a justice that is always yet to come – evokes a radical trangressivity that is finally beyond the disciplinary reach of normative paradigms. In taking up the problematic of some recent conjoined twins as the limit case of the disabled body, I question whether such highly abstract theory can yield new ways of thinking otherwise about those forms of anomalous embodiment that frustrate social, cultural, and legal normativities.

'Queer Pleasures' continues the project of thinking otherwise, and of how it might be possible to exceed the socio-cultural normativities of sexuality in a productive way. For all that Foucault signals some instances where 'bodies and pleasures' might subvert normative stability, his alternative course is underdeveloped and he remains clearest in setting out the disciplinary techniques aimed at the sexual pleasures of the singular body. Though he insists that sexuality is always in a state of dynamic process, neither fully open to, nor limited by, intentional possibilities, it always remains subject to external regulation, 'organized by power' (Foucault 1979a). But to theorise sexual expression only according to such a model of governmentality, or indeed to the alternative psychoanalytic model proposed in Chapter 4, gives an insufficient account of how the performativity of sexuality might be queered.

What is at stake here is not the instantiation of a sexual subject as such, but rather the performitivity of desire – which in Deleuzian terms has an extended meaning – as an element of *self-becoming* that infuses all aspects of the materiality of living in the world. In enlisting a Deleuzian analysis, the chapter traces out a more affirmative account of disability and sexuality that takes up a queer reading of desire as a produc-tive positivity that overcomes the binary of normal and abnormal and celebrates the energies and intensities of multiple transformatory con-junctions. Where other models are engaged with the – albeit contested – boundaries of the embodied self, Deleuze and Guattari propose an extensiveness that forgoes conventional distinctions between whole and 'broken' bodies, and between the organic and non-organic. In conse-quence neither the disabled body in general nor the prostheticised body is excluded from discourses of desire. Their approach signals a break with any stable sexual identity, indicating instead that subjectivity only emerges through the erotics of connection.

That Deleuzian insight continues into Chapter 7, which further enlists queer theory to productively rethink the issue of disability in the con-text of global corporealities. Taking up both the phenomenological and Deleuzian theme of the profound interconnectivity of embodied social relations, the chapter traces out how the spacio-temporal transforma-tions of globalisation are felt in the instantiation of disability, and how disability in turn might exemplify the deterritorialisation that global movements imply. Against the view that globalisation further disadvan-tages those who are already excluded, I suggest – with some caution – that the work of Deleuze and Guattari could provide a potentially positive understanding of the relationship between globalisation and disability. The unpredictable energies and flows of desire already dis-cussed in Chapter 6 with regard to a reconceived sexuality are just as much in play on the macro-level. With reference to Haraway's mobil-isation of the networked cyborg which contests the embodied fixity of the human subject, and in the light of the postmodern technolog-ical possibilities of bio-communications, virtual reality and transgenic species, I speculate on what would come about if the machinic link-age of intercorporeality were to be positioned as the basis of a queer global ethics. In the era of the posthuman, it is unlikely that mere cor-poreal variation could be advanced as a justification for exclusion. That and other bioethical themes are developed in the concluding chapter where I reiterate that flux and instability are not peculiar to the anoma-lous body but are the conditions of all corporeality to the extent that the wholeness and integrity of the embodied subject is a phantasmatic structure. As such, instead of understanding the 'disabled' body as an aberration, disqualified from the full attributes of subjectivity, we should revalue it as a variable mode of becoming. If corporeal integrity is no longer privileged, but is shown to be a contested and impermanent mode of embodiment, then the grounding of forms of subjectivity and sexuality in the dis/abled body is no longer contentious. Indeed, if post-modern conditions of living are changing so radically for everyone, then the modernist anxieties around non-normative embodiment are already a nostalgic throwback. Finally I suggest that an ethics based on the distinction between bodies is inadequate, and that a better model would encompass both a phenomenological approach that entails an ethics of contact and touch, and a Deleuzian attention to the flows of connectivity.

In conclusion, I should like to firmly situate this project within the current state of development in critical disability studies, which, as I have indicated, is used increasingly by those who position their own research within the context of postconventional theory in general, and for whom the original challenge of the social model of disability no longer provides an effectively dynamic model. One important dif-ference that marks out critical disability theory from its predecessors is its willingness to step aside from the claim that disability stud-ies not only is, but should be, territory occupied primarily, if not solely, by those who live with a disability. What has become clearer is that as the ground for a form of oppression widely perpetuated in western societies, disability – like alternatives to heterosexuality, or racial differences – poses probing questions about the nature of those societies, both in terms of their overt organisation and their social imaginaries. The responsibility for enquiry and analysis falls, then, not on disabled people alone but on all those who partici-pate in the relevant structures. Just as the scholarship of recent years has identified racism as a problem of whiteness, so too ableism must be addressed by both those who are identified with normative stan-dards and those who are excessive to them. By taking on a range of contemporary critical theories and asking how their deployment might shed further light on both the operations of, and the reasons behind, the othering of disabled people, no one perspective is privi-leged above others. Whole new areas, like that of sexuality, which had been previously sidelined as politically inessential have been opened up to scrutiny, and deconstructive enquiry has been directed inward as well as engaging with material realities. The key to the new schol-arship is critique, not in the sense of a one-directional destruction of old certainties, but rather as a risky enterprise that subjects all the conventions to new and potentially disruptive analysis. As Judith Butler notes:

> What [critique is] really about is opening up the possibility of questioning what our assumptions are and somehow encouraging us to live in the anxiety of that questioning without closing it down too quickly. Of course, it's not for the sake of anxiety that one should do it...but because anxiety accompanies something like the witnessing of new possibilities.
>
> (quoted in Salih and Butler 2004: 331)

The coming together of disability, subjectivity, and sexuality under the gaze of postconventional critique will certainly generate dangerous discourses, but finally it is a necessary move of ethical responsibility.

1
Corporealities

The project of theorising disability, of enquiring into its status as a dangerous discourse, throws up from the start some intriguing conundrums. On the one hand it is the site where body theory, that most vibrant of recent academic pursuits, has made some substantial and highly creative advances, and on the other it is still widely and dismissively seen as a minority concern of real interest only to those who are themselves disabled. It may certainly be the case at present that the majority of scholars within the field live with a disability, but my own passionate contention is that what the project tells us about being human is of high significance, both to every one of us on a personal level, and across the disciplines. A similar point could undoubtedly be made about feminist theory, postcolonial studies, or queer theory, those other areas of enquiry that deeply unsettle personal and academic conventions. What follows in the case of disability is that the object of study is simultaneously one that attracts widespread disinterest and even derision and yet has the capacity to change the way in which we think about human embodiment in the contemporary world. When David Mitchell and Sharon Snyder suggest that disability as the figuration of materially unfit bodies can be understood 'as the master trope of human disqualification' (2000: 3), their rhetoric is well-supported by the texts they address. That gloomy assessment, however, is rapidly offset by an acute appreciation that what is at issue is not some inherently negative state of being, but one that crystallises the vulnerability of embodiment in general. As Michael Bérubé succinctly notes, 'the instability of disability (is) a device for destabilising all categories of identity' (2002: x). That realisation is the ground for an ensuing tension between the implicit fears that would silence or evade disability and the optimistic hope for change that is not about ending the multiple insults to disabled people – although that would be a by-product – but about opening up the discourse to the very instability that disability embodies. My purpose here in addressing the central concern of corporeality is to begin that process by pushing the terms of the debate along some less-travelled, but potentially productive, paths.

For all the common usage of disability as a universal category, the multiple differences and variations that it encompasses are usually overlooked in the interests of maintaining what is taken to be a primary distinction between bodies themselves: those that conform to normative parameters and those that do not. Certainly many disabled *bodies* do transgress the morphological conventions of western modernity and thereby disturb entrenched socio-cultural determinations of proper bodily form, but that is too limited in scope to explain the anxiety that disability evokes. I prefer instead to widen the terms of reference to the mode of corporeality which accommodates not simply the materiality of the body, but the manner in which it is experienced and lived by an embodied subject. Where visible appearance remains the privileged determinant of what it is to be disabled (although it may in fact disconcertingly offer no indication of difference), the notion of corporeality speaks to the instantiation of subjectivity itself, where – in postmodernist accounts at least – binary thinking is far more difficult to sustain. It is not that the visual can be superseded for it continues to underpin all manner of damaging modernist assumptions, but that any analysis of the relation between disabled and non-disabled must attend to less overt points of anxiety. The discursive construction and maintenance of the ubiquitous stereotypes of disability that effect a devaluation of some specific aspects of bodily difference, and thus of many disabled people themselves, are displaced in contemporary theories of embodiment that uncover instead the uncertain nature of *all* embodied selfhood. And at the same time that the modernist focus on the boundaries of bodily difference gives way to a fuller recognition of the dimensions of embodiment – which is taken to include affect, emotion, and not least, sexuality – corporeality itself segues into the notion of intercorporeality. In going beyond the constitution of a supposedly secure and relatively fixed subjectivity – instantiated through the exclusionary strategies of bodily abjection, separation, and distinction – the insights of phenomenology suggest how the hierarchy of the ablebodied/disabled binary that marks out which bodies are to matter might be productively disrupted by a perception of being-in-the-world-with-others. In such a relation between self and other, where the viability of the atomistic and self-referential experience of each individual is put in question, the mediating factor is often, though never exclusively, touch. I shall return shortly to that aspect, but want first to recall briefly how the normative economy of self and other functions within western discourse.

To be a self – and more significantly a subject – with effective agency is, in every sense of the word, to be capable of exercising autonomy. Because the western logos privileges the freedom and rationality of the putatively disembodied mind as the irreducible marker of the sovereign subject, it follows that the body itself is relatively unimportant, so long as it falls within normative parameters that pose no hindrance to the implementation of self-determination. In short the ideal body's neutrality with respect to autonomy is assured by its normativity in terms of morphological form and function. Once the body is deemed to fall short, however, or suffers some loss of function, it attracts unwelcome attention and becomes the locus of a devaluation that extends far beyond the materiality of any real or perceived deficit. The whole range of emotional and social meaning and significance that is directed to agency and autonomy becomes fully apparent not in relation to the body that passes as normative, but in the response to impairment and disability. What may be taken for granted in non-disabled subjects – the assumption of self-sovereignty – becomes a matter of doubt or denial in the matter of the differently embodied subject. It is perhaps no chance contingency that the wheelchair is taken to symbolise disability universally, for as Lisa Cartwright and Brian Goldfarb note: 'Purposeful mobility, like speech and gesture, is a key signifier of human agency and personal expression' (2006: 139–40). One might think, too, of the biblical marker of the human as opposed to the mere animal as upright, erect, and those ubiquitous images of the evolutionary process that depict the move from the swamp-dwelling reptilian form of early life, through progressive stages of crawling on all fours to the final achievement of standing straight-backed on one's own two legs. The attainment of independent being is clearly linked to control over the body, which having apparently reached its developmental zenith remains a static and relatively insignificant entity unless beset by disease or disability. Yet the focus of attention here – the physical body – is in neither case coincident with the phenomenal body as it is lived in all its rich and varied experience. And what the turn to phenomenology insists on is that body and mind are always inseparable: that corporeal changes are inextricably reflected in changes to the embodied subject, and moreover that embodiment is a matter of process for every one of us.

Returning to the convention, a further and equally important feature is that both the self and the corresponding other should be fully
independent, closed and secure in his or her own integrity, and invulnerable to extraneous influences. In consequence the proper relation
between the two is characterised as one between equal and autonomous
agents whose coming together is negotiated through implicit contract –
a setting out of rights and duties for example – into which each may
freely enter without compromising their own independence. To be a
subject in a world of others, even to attain full personhood, devolves,
then, on maintaining the interval of distinction. The problem is that
although this familiar model is the cornerstone of liberal humanism,
it clearly does not correspond to the everyday reality of most encounters between self and other. Not only does it ignore the operation of
asymmetrical power, but it elides all forms of experiential difference,
and fails to acknowledge the significance of differential embodiment.
Whenever corporeality does intrude into the self–other relation, as it
surely must, the one whose embodiment escapes the putative neutrality of the normative is seen as potentially compromised, as less
than self-complete, and not to be counted as a full subject. In such
a scenario – in any instance in which someone is defined by a form
of anomalous embodiment – she ceases to be an equal, and becomes
the lesser term in a hierarchical binary in which the unmarked self is
dominant. Nonetheless, that dominance is maintained only at a cost,
not only to the devalued other, but to the one who appears secure in
her personhood. Whether it is couched as postconventional theory or
as a matter of everyday experience, it is clear that the inherent instabilities of the body always threaten to disrupt the possibility of any
fixed relation between self and other. What strategies then must be
in place in order to ensure an illusory security, and how may they be
contested?

What is at stake in the determined defence of the ideal integrity of the
embodied subject is that difference should be held at bay lest it mark the
subject with its own shortcomings. As I have argued in more detail elsewhere (2002), such an impossible task inevitably generates ontological
anxiety that can be alleviated only by disavowing any body that fails
to display the attributes of self-control, and predictability. In failing to
reproduce the ideal image of corporeal invulnerability, disabled bodies
are not positioned, however, as *dis*empowered; on the contrary they
signal threat and danger insofar as they undermine any belief in the stability and consistency of bodies in general. Paradigmatically, such bodies
elicit anxiety for they remind the others of their own vulnerability

and precariousness. The feminist philosopher, Susan Wendell, who has chronic fatigue syndrome, makes the point:

> If we tell people about our pain, for example, we remind them of the existence of pain, the imperfection and fragility of the body, the possibility of their own pain, the inevitability of it.... They may want to believe they are not like us, not vulnerable to this; if so they will cling to our differences, and we will become 'the Others'.
>
> (1996: 91–2)

The self-protective desire for mastery in relation to those others has different guises that may appear far apart or even opposed, but it is not clear to me that any distinction can be made between mastery as *power over* another, and mastery as a *defence against* anxiety. Whatever the surface logic, it is the motivation of the desire that concerns me and that marks the variant operations as two sides of the same coin. There are further complications, however, for if one way to deal with anxiety is by the exercise of power and control, that undisguised mode of domination is matched by the more subtle rubric of protection or benign concern for others. Although the latter may be superficially preferable, a lot of what is justified as caring behaviour tends towards limiting the autonomy of those with disabilities. Taking responsibility for another, even claiming empathy, is rarely straightforward, but potentially masks an unwillingness to engage person to person. And at the other end of the spectrum, in the worst cases, the vulnerability elicited in those who would see themselves as 'normal' can invoke a *dis*engagement expressed as real violence: Margaret Kennedy (1996), for example, claims that people with disabilities are up to four times more likely to experience sexual abuse than the general population. In short, although the evident operation of power 'over' should never be discounted, it invariably masks a more fundamental insecurity that could just as easily result in disavowal.

Most scholars of disability are acutely aware that the damaging consequences of the imperfectly hidden insecurity associated with able-bodiedness may manifest in a need for mastery over the supposed threat of disability to the normative order; what is less well-recognised is that it may take the form of a denial that calls to mind Julia Kristeva's account of abjection (1982).[1] Having in previous work (Shildrick 2002) explored in some detail the link between abjection and the monstrous as an umbrella category that certainly encompasses disability, my remarks here will be brief. Suffice it to say that what the subject must abject in order to secure her own being is some part of herself that cannot be owned. Once disavowed, that element – the abject itself – does not become the oppositional other as such but remains in a suspended location between self and other, posing always the risk of a return that would disrupt the originary subject. In Kristevan terms, any form of anomalous embodiment – and particularly that which overtly contests the discursive ideal – is highly productive of anxiety, insofar as it threatens to overflow the boundaries of 'the self's clean and proper body' (Kristeva 1982: 71). The specific and semi-conscious fear that haunts the subject – which I have identified with a socio-cultural and psychic horror of, and fascination for, the monstrous (Shildrick 2002) – is that the extra-ordinary body's putative lack of self-containment, and its failure to occupy a delimited space, signal the disturbing possibility of engulfment and contamination. The normative subject, in other words, implicitly fears that it will be overwhelmed. As Kristeva's concept of the abject makes clear, the issue is not so much that the body of the other is horrifying in and of itself, but rather that it might infiltrate the space of my own body and effect the very transformations that would unsettle my claim to autonomous selfhood. For all that our embodied subjectivity depends on a phenomenological interface with a world of others, there remains, then, a powerful desire for, and expectation of, clearly delineated bodily limits and boundaries. In seeking to maintain the self-possession and control that the modernist ideal of embodiment demands, we must continually make the distinctions – between self and other, and between categories of others – that enact both psychic and cultural assessments of safety and danger. Given that no interaction is entirely without risk to our fragile sense of self, the relations between self and other operate largely within a scopic economy that privileges the spacing – the interval – of separation.

In that regard, it is perhaps significant that the medical model of disability – which promotes an image of individual pathology – should have held sway for so many decades. Against the supposition that the privileged authority of biomedicine simply overrode other accounts, an alternative explanation could be that it served to both legitimise, and settle, socio-cultural anxieties about the disruptive potential of disabled bodies.[2] Even in those contemporary clinical encounters which are hands-on, there is a lingering anxiety about any breach in the implicit *cordon-sanitaire* that surrounds and is assumed to protect the subject. Healthcare professionals are expected to maintain control and to recognise vulnerability only as a quality of the other. Moreover – and here the interrelated roots of psychic recoil from the abject and the avoidance of touch, as an aspect of phenomenological intercorporeality, become clear – the bodies of the healthcare providers are supposedly irrelevant to their practice. Although many healthcare interactions are ones in which touch may be an integral part, the apparent risks to the dominant subject continue to underlie a characteristic refusal to recognise the mutually engaging existential status of both parties. Yet where the authorised narrative may suggest that the procedure is one-way, the clinical examination is always in fact 'contiguous and therefore ambiguous: bodies that touch are also touched' (Diprose 1998: 37). In the biomedical encounter – at least within the western allopathic tradition – touch is most often intended as utilitarian and non-affective, belying its expressive nature as a mode of intersubjectivity; and with some notable exceptions that are more likely to adhere to nursing than physician care (see Dongen and Elema 2001), it is not about bringing bodies into proximity but about attempting to establish a distinction that holds them apart. Nonetheless, it is my contention that regardless of the power relations that are supposed to manage bodily contact within acceptable parameters, the encounter – which I take as a template for all interpersonal contact – disrupts the usual notion of subject/object that marks the distinction between doctor and patient. Instead neither body is closed, or self-contained, but becomes open to the world of the other.[3]

What is at issue, then, is the anxiety occasioned when the safe distance enacted by the scopic economy gives way to actual or anticipated physical contact. Above all anxiety demonstrates the continual failure of autonomy and self-sovereignty to adequately inscribe embodiment. That the body is porous and excessive to its own boundaries is unsettling in its own terms, but the phenomenological model goes further in showing how the intercorporeal encounter is about a mutual crossing of boundaries that enacts the very means through which embodied subjects are both constituted, and undone. To touch another is in some sense always to compromise control, for even where the intent is outward – whether aggressive or palliatory – we are also touched in return. It is an undecidable moment of exchange, a transgression of corporeal boundaries, that troubles the dimensions of the embodied self for all participants. The point is not that such contacts are necessarily unwelcome – after all, affective life in particular depends on the closeness of others – but that they are openings which inevitably risk vulnerability. Certain forms of affectivity are clearly more risky than others, and in particular, as the point at which the body's lack of closure and self-sufficiency is powerfully displayed, sexuality is a highly significant site at which the constitution of the embodied subject is fraught with both pleasure and danger. To minimise such risks at the substantive level, we are wary of physical contact in general, and have well-developed strategies of avoidance when faced with the unpredictable or putatively dangerous. At the same time, the psychic anxiety implicit in any actual corporeal interaction must be alleviated: it may be condensed and projected outward on to those abject others whose bodily disconformity unsettles the ideal of embodied selfhood. In other words, the socio-cultural mores that conventionally regulate tactile exchange intersect with the interiority of psychic investments.

The discursive association of anxiety and psychic investments will be more fully developed in subsequent chapters, but for now I want to focus on some of the more substantive phenomenological aspects of corporeal interaction. To utilise the work of Maurice Merleau-Ponty to think through the meaning of embodiment and its effects on the economy of self and other is to provide a bridge that links theoretical insight to specific practices. At the heart of the issue is a radical critique of the normative dynamics of the self–other link that seeks to jettison separation and distinction in favour of an *intercorporeality* across differences in which the privileged mediator is touch. The question, then, of how to formulate – without falling back into modernist notions of atomistic individuals – the relation between self and other when a real geographic distance intervenes is not inconsiderable, and will be addressed in due course through feminist philosopher Ros Diprose's notion of corporeal generosity (2002).[4] Nonetheless, even in an initial exploration, it is clear that touch is not limited to the terms of its physicality – the skin on skin contact – but extends into the mode of 'being in touch' rather than separate and distant, and 'being touched' in the sense of emotionally moved. In other words, touch is already an interweaving of mental and material attributes. More importantly, it is, in both its literal and metaphoric sense, that which brings the embodied subject into being. The phenomenological mode that I favour here, and which is becoming increasingly influential in critical disability studies as a counter to the givenness of the body and of the subject in the social model, intends far more than a simple recovery of the hitherto neglected notion of embodiment as an irreducible element of all being. Where that initial move merely rewrites the privileged relation of self–other in a form that relies on the mapping of the boundaries between *embodied* selves and *embodied* others that preserves difference and distinction, a more complex approach introduces the notion of intercorporeality as the fundamental structure of being-in-the-world – or rather, as I prefer to put it, becoming-in-the-world. It reinstates both intentionality and subjectivity to corporeality itself, and understands sensibility as 'the interface of mind, body and the world' (Corker 2001: 40). Where Corker is relaxed about exploring the insights of poststructuralism, a slightly earlier article by Paterson and Hughes (1999) makes similar points about agency, sentience, and sensibility, but engages with a more clearly phenomenological approach to counter what they see as the (then) 'arid materialism of disability studies' (1999: 599). The paper has been an influential beacon, perhaps in part because they indicate an intention to mobilise 'an emancipatory politics of identity' (598) and endorse a 'quest for citizenship' (604). For my own part such terms are perhaps too firmly rooted in modernist paradigms, and I prefer always to explore the overlaps between phenomenology and poststructuralist/postmodernist approaches.

In introducing the notion of 'becoming' rather than 'being', my more general project is not only to reclaim the body, but to uncover its inherent fluidity and lack of completion. Becoming signifies a process that shifts and flows just as the body itself undergoes changes and modifications, not in the sense of wholly foreseeable developments over a life course, but as the irregular and contingent transformations and reversals that unsettle subjectivity – and identity – itself. It is not then that the bodies of disabled people are especially, still less uniquely, untrustworthy in their refusal to conform to normative expectations, but that any mode of corporeality just is porous and provisional. Moreover, as Merleau-Ponty's approach indicates, the phenomenological understanding is that the biological, social, and discursive bodies are equally unfixed and mutually constitutive. None is complete in itself, nor operates independently of the other. How we come to know both ourselves and others, and the world itself is a matter of material engagement, often through the direct contact of flesh and blood encounters that do not simply *affect* us at a surface level but *effect* the very constitution of embodied becoming. What this implies is a significant shift in epistemological and ontological conventions that insist on the traditional split between subject/object and self/other. Instead, the body and the world are inseparable, and the sovereign self of postEnlightenment thought is not only decentred, but rendered unintelligible. In place of a claim to independent subjectivity, I recognise myself as, at very least, sharing an 'intermundane space' with other putative subjects engaging in the world through their own specific bodies. And it is precisely this immersion of our own bodies in what Merleau-Ponty calls the flesh of the world, and our interweaving with other bodies, that actualises our social and personal identities. The flesh of the world implies both the viscerality of our environment – we are of it rather than in it – and a fundamental unity of existence. What might be called the matter of living-in-the-world-in-our-bodies-with-others belies the closure of the subject, and makes clear that I must rely on 'other landscapes besides my own' (Merleau-Ponty 1968: 141). Instead of multiple, but separate and discrete, corporealities, then, there is a tissue of intercorporeality in which each body is open to and affected by the others. Accordingly, our lived experience with others is the basis of our being (or becoming)-in-the-world at all, and the autonomy and sovereignty of the subject are continually undone, even as they are enacted, by intercorporeal encounters.

On a less abstract level, it is plain to see how bodies are constituted and fashioned through contact with those around us, those with whom we are in touch, both literally and metaphorically. As Janet Price writes:

> Partners start to mimic each other's gestures or patterns of speech; friends learn the response to hugs and the touches of hands, carried out in a smooth choreography, which stutters and perhaps fails when meeting strangers. We daily remake the ways we move through public spaces: the young woman delighting in her body but wary of the gaze of workmen on the building site, unconsciously shifting her gait; the wheelchair user changing her speed and style as she scoots past curious children in their pushchairs; the older woman with osteoporosis who walks tentatively through the bustle of commuters, wielding her stick like an offensive weapon.
>
> (Shildrick and Price 2005–2006: ¶9)

All speak to a phenomenological mode of embodiment in which it is through the visual, tactile, and aural contact between flesh-and-blood bodies that we both perceive and are perceived by others. As Ros Diprose puts it: 'it is because my body is given to others and vice versa that I exist as a social being' (2002: 54). One consequence of this way of understanding embodied lives is that the binary division that would separate the categories of disabled and non-disabled makes little sense. We are affected by all those whom we encounter regardless of their sameness or difference. When I, as a putatively ablebodied woman, push a friend's wheelchair and develop my own arm muscles, more clearly articulate my words when meeting with deaf colleagues, or verbally describe the visuals of my presentation to a mixed audience, I am modifying both my bodily comportment and my sense of being-in-the-world. What may start as conscious adaptations quickly become habituations that speak to my embodied subjectivity which cannot be extracted from the relationships and connections I engage in. It is not that any change will solidify and remain static; on the contrary, we constantly remake ourselves, fashioning new forms of self-perception and performance. And although the notion of performativity does not escape constraint – I can run for a bus where someone less mobile cannot, but I am at a loss to communicate effectively with a group of sign-language users – it does make clear that our forms of embodiment are dynamic and to a strong degree not simply other-responsive, but other-constructed. There is no place in this model for either the atomistic subject of modernist thought, or for the belief that some forms of embodiment are more settled and unified than others.

Merleau-Ponty's own concentration is firmly on the phenomenology of human-to-human interaction, but where he uses the term 'the flesh of the world', he opens up, perhaps, even more possibilities than he intended. Certainly if intercorporeality in the phenomenological mode engages with multiple differences, then there is no reason that its application should be limited to the intermeshing of *human* persons. One of the most engaging – touching – accounts of becoming as a process dependent on the other is given by Rod Michalko (1999) in his reflections on his interrelationship with his guide dog, Smokie. Although Michalko does not speak directly of becoming, his account exemplifies precisely how the mutually constitutive nature of all intercorporeality radically contests a number of entrenched conventions peculiar to the binary structure of modernist thought. He writes:

> Whatever Smokie and I do, whatever kind of life we experience together and whatever else we mean to each other, we are 'person and dog' sharing a life together. We are 'human and animal' living in the world and moving through it together. Smokie's presence in my life has reminded me that 'nature' is as much a cultural construction as 'blindness' is, and that distinctions like human/animal, society/nature, nature/nurture are themselves human inventions.
>
> (1999: 9)

In normative terms, guide or assistance dogs are simply functional prosthetic devices that compensate for a loss or lack of sightedness, and the success of the relationship between dog and owner would devolve not on the smooth interlocking of mutual dependency, but on the extent to which autonomy was restored to the human partner. But the very intimacy of Michalko and Smokie belies the promotion of sovereignty, both *of* the human self and *over* the animal, and speaks instead to an affirmation of becoming-together that goes beyond bounded bodies. It might even be said that a phenomenological understanding of the intercorporeal connection swiftly shades into something even more challenging to normative conventions. To a Deleuzian, the intermeshed mode of life signals just that form of assemblage that circumvents the blockage of rigid subjectivity and opens on to productive new ways of becoming in the world. The step from Merleau-Ponty to Deleuze is by no means self-evident, but as I shall suggest in later chapters, the perspective of the latter offers equally exciting, and more importantly viable, challenges to the modernist devaluation of anomalous bodies, whilst greatly extending the possibilities of embodiment whatever form it might take. For present, however, I want to take up further aspects of the phenomenological approach that are deeply significant to the rethinking of the meaning of disability and its affective states.

Given my interlocked themes, it is instructive first to review how sexuality adheres with subjectivity and disability within modernist thought. As I have indicated, the very notion of disability alone seems to be a dangerous insult to subjectivity in general, but it is considerably more damaging when linked to sexuality. Given that most forms of sexual expression speak to intercorporeality, they already – at a psychic level – carry a threat to the autonomy of the normative subject. What is at stake is a quasi-intentional merging of bodies and agency, a transmission of what would in other contexts count as abject matter, an openness of the self to the other that invites vulnerability, and an encounter that is unpredictable and indeterminate. As Michael Warner notes: 'sex is a disgrace... the possibility of abject shame is never entirely out of the picture' (1999: 2). That a sexual partner should be disabled merely serves to amplify the putative danger to the supposedly sovereign self. To that uncomfortable scenario, phenomenology brings new insights and perhaps initially a heightened anxiety. In a general and somewhat abstract sense, the basis of the phenomenological model of embodied subjectivity speaks not to a settled self but to a lifelong process of becoming *with* others in an open encounter that constitutes both self and other. In turning to the context of sexual relationality, however, Merleau-Ponty's notion of intercorporeality and, more specifically, the reversibility of touch (1968), takes on a concrete materiality that figures both an indistinction between the limits of one body and another, and, thereby, the potential for the transfer of impurity. It is for these reasons that devolve on the putative risk of losing autonomy, self-definition, and corporeal integrity that the domain of the sexual is so highly policed – both literally and metaphorically – and so overdetermined by performative constraints.[5] It is scarcely a surprise, then, that where the embodiment of the other is dis-ordered and inherently challenging with respect to the regularities of normative corporeality and its outward affect, that the moment of contact, the touch of flesh on flesh, should indicate not simply the need for self-interested caution, but an intimation of danger, amounting even to self-dissolution. As exemplars of non-normativity, those with disabilities cannot but disrupt and contest the modernist subject's illusion of purity, containment, and self-control.

Although his explicit remarks are relatively limited, Merleau-Ponty's own approach to what he calls 'the body in its sexual being' (1962) already exceeds the traditional purview of philosophy. In elaborating what he means by the subject's 'opening out upon another' (1962: 194), Merleau-Ponty claims that the phenomenon is found everywhere, but most specifically within the development of sexuality. As he understands it, our mode of being in the world – what I am designating 'becoming' – is projected through the sexed body to the extent that sexuality is a modality of existence (Merleau-Ponty 1964). As Judith Butler (1989) has noted in an early essay, Merleau-Ponty's notion of sexuality makes a break with the Freudian theory of drives and instincts, and positions it as 'co-extensive' (Merleau-Ponty 1962: 196) with life itself. The incarnation of the subject is inextricable from its sexuality. What is somewhat troubling, then, is that when it comes to considering contexts beyond the mainstream, it becomes clear that Merleau-Ponty has a distinctly normative – and, as Butler fully recognises, masculinist – approach to what constitutes sexuality. In his exposition of the case of Schneider – a man who had suffered a wound to the head occasioning a degree of brain damage that seemingly obviated his sexual desire – Merleau-Ponty both reiterates the rupture between sexual responsiveness and instinct, referring instead to an erotic perceptual schema, and fails to recognise that what is lacking in Schneider represents a very limited view of sexual subjectivity. Just as feminists such as Butler (1990) and Grosz (1994a) stress the failure to accommodate the differential sexuality of gendered bodies, disability theorists will find little to develop the specific experiences of that other category of morphological difference, disabled people. In tacitly valorising just one mode of corporeality and sexual becoming, Merleau-Ponty leaves too much unaddressed. As Liz Grosz puts it: 'The question of what other types of human experience, what other modalities of perception, what other relations, subjects may have with objects is not, cannot be, raised in the terms he develops' (1994a: 110). Nonetheless, what may be taken from Merleau-Ponty and elaborated is his insight into the non-biological aspects of sexuality, the perceptual opening of one body to another, and an affective sensibility that communicates between bodies, transforming and transposing behaviours, intentionalities, and sensitivities. The sexuality of the embodied subject is no longer a matter of internal drives, but of mutual becomings.

In his final and unfinished work, *The Visible and the Invisible* (1968), Merleau-Ponty greatly develops his understanding of the reversibility of the sensations that flow between us. As he sees it, the unity of our mutual existence is woven together by the reversibility of such binaries as perceiver/perceived and subject/object. In other words, my own expressive body is not just interactive, but finds itself in a chiasmatic relationship with other bodies to the extent that as I see and touch, I feel myself being seen and touched. I experience a reversibility such that, he claims, 'the world of each opens upon that of the other' (1968: 141), and it figures what Liz Grosz calls: 'being as reversibility..., being's dual orientation inward and outward, being's openness, its reflexivity...' (1994a: 44). Put another way, it is not that my existing body is caught up in an exchange of sensation as that the exchange itself constitutes my mode of embodiment. Moreover, given the inherent reversibility of sensation, it is difficult to distinguish clearly between the active and passive mode. To take the example of touch, it is clear that the hand is both sentient (able to feel the other) and sensible (able to be felt by the other), and insofar as touching/being touched is simultaneously both continuous and differentiated, a certain ambiguity always remains. What this suggests is that the boundaries of the embodied subject are highly uncertain, but while Merleau-Ponty does implicitly contest the viability of binary structure, he does not entirely leave it behind, preferring to maintain a pragmatic distinction between self and other. Nonetheless, his account of touch at very least frustrates the hierarchy of subject and object, and shows the experiencing body crossing boundaries rather than creating distance and division. In any case, the grammatical structure of our language is such that it is extremely difficult to avoid terms such as self and other, and perhaps it is always best to imagine them in the phenomenological form, just as in fully poststructuralist accounts, as under erasure.

Although touch is rarely given the same privilege as sight, developmental psychologists (Montagu 1971; Anzieu 1989) tell a different story. Even before birth, in the foetal stage, touch is the primal sense and it remains paramount throughout early infancy. It is only when the growing child enters into symbolic relations as an essentially singular, unified and bounded subject that touch loses its privilege in relation to other senses, and that sight in particular becomes the most valued attribute.[6] By adult years, most people have become extremely cautious about the circumstances in which touch is acceptable, precisely, I would surmise, because an unwelcome or unexpected touch seems to threaten both the integrity of the self, and the ontological separation of self and other. So powerful are the prohibitions of unconsented touching that in law, it constitutes the tort of battery. If touch always has the capacity to evoke unease, that anxiety is noticeably heightened in some very specific contexts. It is especially provoked, for example, by adults with developmental disabilities in that they may hug and touch unselfconsciously and 'inappropriately', where that word intends precisely a contestation of the boundaries that are proper to one's own body. They may explicitly cross the interval of separation that marks out the protected domain of the subject. Staring does something similar, but its violation of the proper never quite erases the distinction between self and other, nor threatens the wholeness of the self. It is significant that in both instances what is deemed improper for the one who is anomalously embodied may be an acceptable mode of behaviour for the one who is normative. It is a short step to the objectifying gaze or the violent or possessive touch that characterises dominant practices like biomedicine. Although the interval of separation may be contested, then, by the possibility of a reversibility that directly relates us to the existence of others, and to our mutual constitution, the relationality of the intercorporeal may be far from unproblematic. In effect, the phenomenologically grounded potential of reversible engagement with the world is undercut by a differential take-up of the possibilities, which relates both to different forms of embodiment, and to the issues of power that may lay behind them. We do not all engage with world on an equal basis, and some of us experience greater restraints on touching than others.

A great deal of work has already been done by feminist scholars (Young 1990; Irigaray 1993; Grosz 1994a; Weiss 1999) in uncovering the lacunae of standard phenomenological theory with respect to sexual difference and in rethinking it to take account of sexed and gendered bodies. The outcome has been deeply productive for feminist theory and strongly suggests that disability studies could benefit in a similar way. What is somewhat surprising is that the take up of phenomenological approaches should be relatively recent there, although the model has already been widely used in relation to the experience of illness. In that context, the critique, and indeed the stepping stone to a way forward, relates to the theoretical claim of mainstream phenomenology that despite the self-evident web of corporeal connections that we all experience, most of the time we exhibit low consciousness (almost a forgetting) of our own bodily engagement, unless, that is, something goes wrong and our bodies are contingently frustrated in their normative functions. That may possibly be plausible for those few people whose bodies rarely deviate from the norm, but it does not fully engage with the notion of intercorporeality as I understand it. At very least, it is clear that the specific bodily competencies of each individual will mean that everyone experiences – or forgets – their corporeality differently, and indeed some are forced into an acute corporeal awareness insofar as their bodies do not coincide with normative morphology. In assuming a norm of inattentive bodily habituation, mainstream phenomenology implies that those who do not seamlessly intermesh with the world as embodied subjects experience bodily discontinuities as interruptions or blockages to their own self-possession. In other words, the body becomes an unwelcome presence that signals limitation and vulnerability. The consequence is that within mainstream phenomenology, issues like illness, ageing, or disability tend to get treated as problems, and are therefore often medicalised in an attempt to re-establish corporeality as controlled and forgettable. The operation of conventional health-care seems directed to once more cover over the presence of the body as uncertain and in flux, and to restore the illusion of self-completion and invulnerability. But this surely cannot fulfil the deeper promise of the phenomenological critique. Not only does it seem to reinstitute a devaluation of certain morphological conditions like physical disability, but it fails to carry through the implications of the phenomenological understanding of the mutual constitution of embodied subjects. By re-exploring two major points of the model, however, it is possible to push it in an alternative and far more productive direction.

First, the classic phenomenological account insists that the embodied self is always in a state of renewal and adjustment in the face of changing physical and environmental circumstances. On that basis, it seems clear that the normative body itself – the fully competent, healthy body – is faced with multiple everyday modifications to its comportment that will require ongoing and intentional maintenance strategies to avoid slippage beyond the norms. In other words, all of us – however we are individually embodied – are more or less conscious that our bodies demand attention. I am not denying that the performative awareness that imposes itself on normative embodiment is substantially different from the awareness of the body that experiences more radical signs of putative breakdown – as in illness, pain, disability, and so on – but the point is that we should not suppose that the embodied self is ever in a state of static integration. On the contrary, we might reflect that it is disarticulation – both actual and potential – that marks every aspect of living-in-the-world. My suggestion is that when we experience anxiety in the face of overt corporeal disorder in the other, we do so not because such disorder is an unknown quality, but precisely because it is always already our own repressed experience of embodiment. In early infancy each one of us has experienced *le corps morcélé* – as Lacan (1977a) names it – the body in bits and pieces, and it is only in the process of becoming a subject in the Symbolic that that image is disavowed.[7] In other words, the bodies that evoke anxiety and a fear of touch – those who are old, or sick, or disabled – are not so much strange as all too familiar. They are abject, but their disarticulation is that of the normative subject. In short, any unsettling phenomenological transformation to an individual's sense of self may cause all those in relationship with her to be brought face to face – according to the strength of their investments in a phantasmatic unity – with their own originary body 'in bits and pieces'.

The second, and related, point is that if phenomenological theory is taken at its word, then the concept of becoming-in-the-world-with-others implies that changes or transformations to the embodied self do not happen in isolation, but mobilise, as it were, a chain effect in the embodiment of contiguous others. Where it is straightforward to see that if I break my own leg, the nature of my corporeal being is quite fundamentally realigned, it is no less clear to me that if someone close to me breaks her leg, then some corporeal readjustment may occur in me. And nor does the modification have to be dramatic: I am suggesting that the same may be true of less radical changes in embodiment. Nonetheless our usual response is to act as though change could be individualised and contained. By way of illustration, I would cite my own long-term collaboration with a friend who has a serious and progressive disability which at first glance might simply position our respective forms of embodiment as profoundly different. Given a mutual commitment to theories of the body that speak to the instability of all corporeality, it has become increasingly apparent to us that the privileged form of disability writing – into which we endeavour to intervene – regularly enacts an occlusion that reiterates a binary split between disabled and non-disabled bodies that is at odds with the phenomenological reading I am proposing. Either a considered standpoint position is adopted, or, moving beyond theory, the characteristic account is an autobiographical one in which what counts is the experience of the disabled person alone.

As Mitchell and Snyder remark: '(a)utobiographical narratives demand that the disabled subject develop a voice that privileges the agency of a bona fide perspective of disability.... a specific and distinct perspective of its own' (1997: 11). What is the case more often than not is that the account of the narrator's own anomalous embodiment fails to recognise or acknowledge that difference and vulnerability are, to a greater or lesser degree, qualities of all of us. Instead such narratives encourage an image of the person with disabilities as not simply having very specific experiences of embodiment – which is incontrovertible – but as being distinctly other in her corporeal specificity. The problematic of bodily difference is situated as an issue for her alone, and the engagement of others, whether read as assistance, or interference and control, is positioned within a quasi-contractual relationship of distance and distinction. Her body is more likely to be acted on, rather than engaged with in terms both of mutual difference *and* commonality. The collaboration between Janet Price and myself has, we hope, taken a more productive approach, specifically because the medically deteriorating condition of one of us has provoked extended reflection on the disorder, vulnerability, and instability at the heart of *all* human becoming.[8] Not only has the illusion of corporeal and psychic wholeness been thoroughly disturbed, but so too has any sense that the embodied subject can maintain separation from its others.

As I have already indicated, our everyday relation to tactility is so far from assured that it is hardly surprising that in many disabling conditions one of the most difficult transactions to renegotiate is that involving touch. It is not just that disabled people may find themselves being touched in a manner that exceeds normative constraint, but that the sense of touch itself may be disrupted. When in the course of a period of acute disability, Janet Price's tactile responses were chaotic and unpredictable, I also lost confidence in tactile interaction. In a subsequent joint paper (Price and Shildrick 2001), we wrote of our shared experiences during that period, with Janet eloquently addressing her own sense of touch and intercorporeality in relation to a number of friends and carers, while I reflected on the unexpected extent of my own uncertainty. As I wrote at the time, having no clear sense of Janet's own corporeal boundaries, I found it difficult to know whether a greeting hug, for example, would be experienced as a sign of affection, would be literally unfelt, or even would constitute a more or less painful assault. In the absence of meaningful response, my own bodily gestures felt clumsy and ineffective. In more regular circumstances, and particularly where a relationship was functional rather than emotionally engaged, we might rarely need to reflect on the mutuality of touch. In the context which I am describing, however, being in touch had wider dimensions, and the corporeal disruption of the one spread to the self-experience and self-understanding of the other's body. Things that were habitually taken for granted – 'forgotten' in phenomenological parlance – became open to question: the interactive bodily skills that once learned are unremarkable and require little conscious thought to enact are revealed as reliant as much on the response of the other as on subjective agency. In specifics here, the disruption to the reversibility of touch troubles the mutual constitution of our embodied selves.

What this brief fragment of autobiography illustrates is that the instability of the disabled body, far from being peculiar to that putative category, is simply a more acute instance of the instability of all bodies. The coming together of anomalous and normative embodiment does not erase the recognition of radical and irreducible difference, but lays bare what is at stake in *every* encounter between self and other. It figures a very different approach to the question of limits and categorical boundaries that goes well beyond the recognition that any body can 'break down', and reveals instead that for everyone, the appearance and experience of corporeal unity is highly contingent and dependent on intercorporeality. In terms of modernist discourse, we may feel compelled to enact our transactions with others under the constraints of normative separation, but it could be otherwise. If the notion of becoming-in-the-world-with-others were taken seriously, the very sense that our own bodies are proper to us and are rightly protected by the interval of distinction, and that autonomous agency is the desired standard for all, would lose its grip. To substantiate an alternative ethic of relationality – that did not rely on the autonomy of the singular, detached and self-complete subject – would go at least some way towards forestalling the anxiety, and even hostility, evoked by proximity. Any encounter is a complex, but rarely acknowledged, mix of uncertainty, emotional and psychic investment (of which I shall say more later), and sensation, to the extent that each has what Ros Diprose calls, the 'structure of an indeterminate, ambiguous relation to the world of the other' (1998: 38). Proximity and touch are never without risk, but the move to deny them, to invoke the putative safety of the interval, is bound to fail. What phenomenology tells us is that we are always already exposed, already immersed in one another, and that in acknowledging our intrinsic openness to the other – all the others – lies the best hope of overcoming the insistent hierarchies that strip some bodies of meaning and value.

The implications of using postconventional and specifically phenomenological approaches to rethink the way in which disability can be understood are potentially very significant and open up new social and ethical possibilities. The dis-integrity and permeability of bodies, the fluctuations and reversibility of touch, the inconsistency of spatial and morphological awareness, the uncertainty of the future are all features that may be experienced with particular force in the disabled body, but they are by no means unique to it. The stress on the instantiation of subjectivity through relationality – even as it remains ambiguous and in process – speaks to a model of the self that cannot be thwarted by disturbances to the body alone. Those who undergo changes to their sensory apparatus, or to their mobility, are undoubtedly changed at a very fundamental level, but they are not thereby in deficit. Within the terms of phenomenology, the embodied self incorporates difference and modification not as a pre-given subject adjusting to evolving constraints and possibilities, but in aligning or realigning one's whole being to whatever perceptual schema is available. As Merleau-Ponty notes, the blind man using a stick to guide his way is not making do in a utilitarian sense but entering into a new relation with his world: 'the stick is no longer an object perceived by a blind man, but an instrument *with* which he perceives. It is a bodily auxiliary, an extension of bodily synthesis' (1962: 176). The prosthetic enables a very different body image to that of a sighted person, but both partake in what Merleau-Ponty calls, a 'total awareness of my posture in the world' (1962: 100). To have a disability, whether congenital or acquired, varies, but does not break, one's immersion in the flesh of the world, and rather than being seen as negative, it might equally signal the opening up of new horizons. Even major trauma – such as spinal cord injury – in which the body undergoes sudden and radical change could be seen as an expansion rather than diminishment of one's becoming-in-the-world. I do not mean to suggest naively that anyone going through such bodily disruption would not experience loss, even despair, in the face of changed circumstances, but that as the new phenomenology of embodiment becomes familiar, different forms of perceptual awareness and interrelationship may become a site of unexpected possibility. The difficulty is that so long as we remain caught up in a modernist view of the world in which the invulnerability of the sovereign subject is paramount, such insights are resisted or degraded.[9] Yet however we are embodied, we remain immersed in the flesh of the world where experiencing and being experienced by others is not a formal encounter between self and other but a matter of intercorporeality. There is no transactional hierarchy.

Having now established the complex dimensions of corporeality in general and its specific significance to disabled people, it should be clearer why the social model of disability has met with increasing criticism. I am not sure that Mike Oliver would any longer repeat his now infamous dictum that disability has nothing to do with the body (1996), but it remains the case that to primarily underscore the external socially situated determinants is to severely limit and constrain what can be said about the experience of disability. As Helen Meekosha remarks: 'A proper focus of social analysis includes subjective experiences of suffering, pain, rejection, loss, grief, desire, joy and achievement' (1998: 167), and she goes on to note: 'Focusing only on the disabling affects of a prejudiced and discriminatory society with a political project geared to changing institutions, beliefs and practices leaves the impaired body as untouched, unchallenged; a taken-for-granted fixed corporeality' (175).[10] There are, then, two important things going on around the notion of corporeality: the first that it grounds and continually interacts with the processes of subjectivity, and second that to absent the body from consideration implies an essentialist understanding of biology that is completely at odds with contemporary body theory. Many feminists in particular have been highly sceptical about any strategy that, in responding to the historical charge that the immanence of the body is supposedly the mark of a femininity unsuited to transcendence, has enacted its own devaluation of women precisely by bracketing out the corporeal dimensions of everyday life. In a move that may remind us of the SMD, early second-wave feminist texts are at very least ambiguous about the significance of embodiment, and see its gendered specifics as an obstacle to social and political liberation. More recent theory, however, has fully embraced corporeality in all its contingency, fluidity and messiness, and instead of trying to fit the feminist cause to the demands of the modernist logos, feminist scholarship has been at the forefront of the challenge to that convention and has insisted that corporeality be rethought.[11] The move has effectively opened up the original political impetus to questions of 'why' rather than simply 'how', and allowed for an extension into an analysis of the psychic structures that operate alongside the traditional conceptions of external power. Something similar is gradually emerging in critical disability theory, and it undoubtedly provides a stronger base from which to understand the interplay of disability, subjectivity, and sexuality. In contrast to the SMD, there is no direct political programme at stake, yet to think together – for that is what phenomenology demands – and to think differently might just change everything.

2
Genealogies

The contestation of received conventions about the nature and embodiment of disability has been greatly advanced by many of the theoretical resources developed during the twentieth century. As the preceding chapter has outlined, the discourse of phenomenology has been particularly influential in opening up alternative ways of understanding the embodied self in general and, like the poststructuralist notion of performativity, has deeply unsettled any notion of a fixed and given subject whose agency may be simply either advanced or hindered by its particular anatomical, physiological, or mental comportment. As with the genealogical analytic that this chapter will develop, neither phenomenology nor performativity was initially addressed to the problematic of disability, but what all three discourses show is that traditional ways of reading the body are inherently open to a process of rethinking. This is especially pertinent to the relationship between 'standard' forms of embodiment and various anomalies, for where normative constructs are often positioned as self-evident and remain unproblematised, irregularity alone attracts attention. If we leave aside the (contestable) claims of biomedicine, it is not that a contemporary or twentieth-century rethinking of the body in all its variation is grounded on a more accurate or sophisticated account of the substance of corporeality, but that the meaning given to bodyliness has been privileged above empirical explanation. Indeed the postmodernist argument is that alongside the multifarious material changes to the morphology and capacities of the body that in themselves generate a need for a new approach,[1] the import of corporeality has always been shot through with contradictions and transformations that undermine any confidence in a fixed and given corpus.

For recent disability scholarship, Foucault's explication of both the emergence of modernist normativities and the operation of power/knowledge has proved fertile ground. His genealogical project, moreover, is explicitly directed to the status of the body, to 'its conditions of weakness and strength, its breakdown and resistances' (Foucault 1977a: 144), and to the recovery not of origins or essential foundations, but of a discontinuous discursive materialisation. His understanding of the relationship between the categories of normal and abnormal, and the way in which difference is equally constructed and regulated has thrown light not only on a medical model of embodiment that pathologises the disabled body, but also on the more recent social model of disability that effects a politicisation of the problematic. Where the former has been as it were imposed from the outside by those who count themselves ablebodied, the latter has emerged from within the disability movement itself, and yet both accounts are deeply implicated in the circulation of power. Alongside the deployment of Foucault's return to the *body* as both support and challenge to the hegemony of sociocultural determinations (an insight that is all too often forgotten in the SMD), I go on to offer a further complication. Where his approach speaks to external discursive power – albeit embedded in the individual consciousness – as the motor for change, I seek to supplement and reconfigure the problematic by engaging with the operation of psychic elements. Within the historically contextualised transformations in the meaning of disability, there is, as I go on to argue, a further set of considerations. Where Foucault focuses his analysis on the multifarious exteriorised mobilisations of self and other, and related binaries, what might be broadly termed a more deconstructive agenda serves to discover the other *within* the self, as an element interior to, and constitutive of, the embodied self. Such a move to the interiority of being, which posits a certain persistence in psychic responses, not only reinforces the contestation of periodisation in conventional accounts, but also implicitly critiques the alternative Foucauldian approach. The trajectory of this chapter, then, will play over some historical moments and bring them together with postmodernist and psychoanalytic theory to demonstrate how the security of binary differences is constantly undermined by the irreducible *différance* – the refusal of the self/other relation – of the disabled body.

As a category of enquiry, disability has a long history of description and explanation that has focused primarily on the condition of being other, on the operation of modes of segregation, and more recently on the *active* silencing or disavowal that paradoxically serves to mark that separation more forcefully. Inspired by Foucault, then, my starting point is to approach the history of the disabled body by seeking to expose how that established narrative may be deeply problematised by a genealogy that disrupts the notion of a continuous and progressive development of ideas and images. As Foucault explains it in one of his lecture from the series given at the Collège de France, genealogies uncover and contest in particular, the 'centralizing power-effects that are bound up with the institutionalisation and workings of any discourse' (2003b: 9). In place of the successive materialisation of ever-more refined epistemologies of disability that build on existing – often scientific – knowledge, the focus shifts to competing discourses, culturally embedded in particular historical periods and locations, and to the notion of discontinuity and fracture. Although certain authorised discourses come to dominate, their mainstream ubiquity is held to be temporary and unstable, despite what may be an extended persistence through time. While any brief account of historical change is inevitably an oversimplification that seems to imply certainty where none exists, the Foucauldian approach, whilst often challenged as to its factual accuracy, is best read as heuristic and indicative rather than as a determinative programme. In providing an approach that links the construction of bodily normativities in particular to the power of the normal/abnormal binary, Foucault's work has been taken up increasingly in disability studies to chart and analyse a historical trajectory that signals at least three distinct models of the disabled body.[2] That the approach has gained a favourable reception is due not least to its potential to support the demand of the disability movement that disabled people should recover their own voices and move away from passivity to a putative self-determination. While it is an exemplary moment in the evocation of what Foucault would call subjugated knowledges, it is not one that should encourage complacency. As Lisa Diedrich points out: 'genealogy helps to locate some of the lost events of disability, *even or especially those events covered over by the emergence of disability studies and the disability rights movement in its current form*' (2005: 653, my emphasis). The task of disability scholarship is never-ending.

The dominant paradigm identified with much of the pre-modern period, then, and indeed one that has not entirely disappeared from the contemporary scene, is that driven by spiritually inspired and superstitiously expressed notions of human perfectibility and a god-given nature. While it is not the case that all forms of disability were always viewed in a negative light – and there is particular ambivalence with regard to those classed as monstrous[3] – there was nonetheless a clear distinction between perfect and imperfect bodies that could manifest in a strong aversion to, even persecution of, disabled people. With the rise of modern science and what Foucault names as biopower, that 'irrational' response to disability was largely superseded by the medical model that still retains a privileged place in contemporary mainstream culture. In being firmly aligned with illness and disease, both physical and developmental disability fell to the remit of medical practitioners where it was treated and contained as a matter of individual pathology. In the last couple of decades, that latter discourse has been increasingly displaced by the social model of disability that reads the imposition of normativities in an entirely different light that seeks to remove any sense of personal stigma and, moreover, to politicise the oppositional relationship between disabled and non-disabled. In place of a privatised and pathologised mode of 'suffering', the conditions of disability are taken to be constructed by environmental, social, and economic factors.[4] And although the newer model has evolved from broadly Marxist underpinnings that would imply some teleological structure in which an underclass might emerge from oppression, it nonetheless demonstrates Foucault's dictum that every mode of resistance, in the process of gaining credibility and authority, becomes caught up in the cycle of power/knowledge. In other words, consistent with the Foucauldian analytic, each of the accounts outlined – including the social model of disability, which is explicitly promoted as a resistant discourse – constructs and polices its own privileged normativities and thereby mobilises new counter-discourses and resistances. In short, no account, whatever its provenance, can claim final authority, but will always generate a challenge to its own limits and legitimacy.

As a heuristic method, the analysis of how western normativities have developed and been consolidated provides a valuable point of entry into any understanding of contemporary disability. Against the dominant standard, the construction of physical difference as a failing, incomplete and inferior, marks disabled embodiment as deeply devalued, not so much for what it is, but for what it fails to be. Its status, value, and meaning are from the start relational, rather than having autonomous standing. Regardless of whether the focus is on the body itself or on the socio-political context, there is broad agreement among analysts that far from being a bioscientific fact, disability is a category constituted, given meaning, and expressed through an endless set of cultural, historical, political, and mythological parameters that ambiguously define disabled people as excessive, as contaminatory, as both malign and helpless (Mitchell and Snyder 1997: 2–3; Shildrick 2002). The otherness of disability is, then, a disordering otherness within the social body that can be contained only by the strict imposition of normative categories that separate out and hold apart the supposedly oppositional groups. In signifying disease, trauma, and decay, the anomalous body is an uncomfortable reminder that the normative, 'healthy', body, despite its appearance of successful self-determination, is highly vulnerable to disruption and breakdown. As such, disability is always the object of institutionalised discourses of control and containment, in which the pretensions of biomedicine to scientific objectivity and rationally based action have been convincingly problematised by Foucault (1973, 1979a) to reveal a series of constructive and regulatory operations. Behind that undeclared disciplinary intent, however, lies another level of signification that plays no part in Foucault's project of laying bare the organisational mechanisms by which both positive and negative identities are constituted and differentially imposed. Alongside an analytic focused on uncovering the operations of the binary relations between self and other as they pertain to disability, and which thereby reinforce the categorical separation in play, it is instructive to pursue a more thoroughly deconstructive approach that may apprehend the disruptive positioning of the other within the same.

My argument is that, whether biomedical or otherwise, responses to disability are never objective, nor adequately analysable in terms of external oppositions, precisely because binary differences are constantly undone by the irreducible operation of *différance*, an imbrication of self and other that frustrates separation and distinction. Self and other are mutually invested, not as complementary entities, but in the degree that otherness is an interior attribute of the embodied self, albeit one necessarily denied by the sovereign subject of modernity. Given the explicit privileging of wholeness, independence and integrity demanded of the ablebodied subject, the cultural imaginary[5] is highly invested in fantasies of an invulnerable body, which in the face of a disability that threatens always to claim its identity in the selfsame produces an anxiety that cannot be allayed. To be 'temporarily ablebodied' is not simply a temporal matter, then, but speaks to the inherent instability of normative embodiment in general. As Hannabach notes: 'After all, the normative body, with its singular gender, race, and sexuality identification, is established only through the active expulsion of embodiments, perceptions, and identifications that exceed narrative and political unity' (2007: 256). In contemporary times, it is easy to mistake the anxiety evoked by disability for a relic of a less sophisticated understanding of the human body, and a harsher enforcement of the differential valuation of what is same and what is different, but that would be to misconstrue its nature. My point, in contradistinction to such a comforting interpretation that sites the operation of disableism as a failure of knowing, is that the psychic content of the anxiety – the threat to the integrity of the self – remains embedded in the present-day domain. Moreover, as Kristeva reminds us, 'Freud always maintained that psychic life was doubly determined – by the biological domain and by the symbolic domain' (1996: 85). In other words, bodies – and in this case, the manifestation of the disabled body – always continue to have a direct impact that goes beyond a changing signification. Psychic anxiety, then, persists even though in recent decades its public expression may have taken less overtly oppressive and discriminatory forms. For all that modernist ethics and politics espouse the values of equality, and more recently diversity, disability is still positioned as the other that not only disturbs normative expectations, but also destabilises self-identity.

Without claiming to give any kind of overview, I want to look briefly at some historical moments in the relationship between disability and society that make clear that what is at stake is far more complex than the variable socio-political attempts to respond adequately to a range of physical differences. We should be cautious in any case about assuming that perceived changes are clear-cut and unambiguous, still less that disability has a stable and progressive history. It is, rather, as Foucault might have put it, a history of *mentalitiés*, in which there have been multiple shifts and reversals in how disability is defined and perceived. But where Foucault has been read as overdetermining the discontinuities between paradigms, it is important to acknowledge that successive trends do not simply supersede what has gone before. Although significant developments and transformations occur in the dominant discourse itself, there is never a single discourse at work, but, as Foucault recognised, a complex mix of interwoven ideas and beliefs that belies the notion of historicist or progressive periodisation. This is both significant to my argument and widely overlooked by many of those working in the field of disability. Even a scholar as nuanced as Rosemarie Garland-Thomson, for example, talks without qualification about a marked shift in the early modern period away from the mode of the marvellous – when disabled bodies were often classed alongside the monstrous – to the mode of the deviant when differential forms of embodiment became increasingly medicalised and pathologised (1996: 3). The inevitable result is a certain flattening out of comprehension that misses a more subtle insight into, and analysis of, the covert psychic operations underlying a materialist surface. Despite the tendency, then, to talk about history in positivist terms, I prefer to see it as concerning the constant circulation and recirculation of ideas – both articulated and hidden – that are intermeshed, one with another. It is not simply that the past can only be known selectively, dependent on what is recorded and preserved, but that we read it through our own representations, our own beliefs and value systems.[6] Moreover, I would liken history to the structure of language, where signifiers are infinitely reinterpretable, and the signified changes constantly with reiteration, such that there can be no question of recovering a pure origin.

In the light of these provisos that nothing can be known with certainty, and that any emergent discourse is contaminated by contemporary partialities, my own analysis must remain open. Although any positivist setting of cause and effect is to be avoided, there are, nonetheless, some tendencies which can be broadly supported by the available evidence, and these will be outlined before I move on to supplement that approach by a more overtly speculative enquiry. In relation to the extant historical material, what emerges in terms of socio-personal responses to disability is best understood as an oscillation between the processes of integration and invalidation or exclusion, threaded through in each case by a persistent and often unacknowledged anxiety. Indeed, it is that transhistorical anxiety – by which I mean not exactly fear but more a sense of unease that finds expression in every period – rather than the material and specific context in which it appears that is the centre of my enquiry. For just these reasons, the naming of discernible historical changes as progress – even where the immediate and substantive conditions of disability have been improved – is fraught with difficulty. I do not mean to suggest that all responses to disability are uniformly negative, but that even the most seemingly benign socio-cultural developments or personal attitudes may merely mask an underlying fear of that which resists the closure of final classification. Moreover, the very real physical violence which continues to erupt against disabled people as a category, and which can be experienced on the individual level at any time, is no mere aberration, but is paralleled by a less obvious discursive violence that is an intrinsic feature of any binary system of sameness and difference.

In *A History of Disability*, the French theorist Henri-Jacques Stiker offers a challenging explanation of why disability – despite multiple contextual variations – is indeed transhistorically disturbing. His claim is that radical differences in individual morphology signify a more threatening disorder. As he puts it, 'an aberrancy within the corporeal order is an aberrancy in the social order' (1999: 40), and more provocatively he indicates an ontological threat: the disabled 'are the tear in our being' (1999: 10). Stiker's approach is far-removed from the positivism that worries me, and from the outset it is clear that his work is as much an implicit deconstruction of contemporary narratives of socio-cultural and political progress as it is an exposition of historical material relating to disability. Written in the early 1980s, the work both predates and anticipates the explosion of disability scholarship around the social model, albeit with a far greater attention given to the discursive construction of the body itself. Although Stiker scarcely references Foucault, his attention to *mentalités* and to the disciplinary practices directed to the body takes on a quasi-Foucauldian approach, but one that is decisively opened up to other considerations.[7] Not least is his very clear and frequently marked awareness of the operation of a cultural imaginary that is deeply invested in normative fantasies. As David Mitchell remarks in his *Foreword* to the book, Stiker 'seeks simultaneously to demonstrate the presence of disability throughout Western history and to unmoor our collective fantasies in the "promise" of eradication or cure' (Stiker 1999: ix). It is an analysis with which my own approach widely overlaps, but alongside the central, though under-regarded, *presence* of disability, I should want to add in a yet more disturbing nomination, disability as an absent presence. The success and/or failure of strategies of both integration and exclusion speak precisely to the *différance* of disability, and to its status as the other within. The disabled body is deeply disruptive to the social body and to normative selfhood alike, not so much in the guise of the clear and distinct other that can be grasped in its difference, but because it remains undecidable, neither self nor other.

That such a notion has evolved from postmodernist theory is immediately apparent, but it should not be limited to an analysis of present-day practices and values. In other words, any examination of the historical trajectory of disability is open equally to a reading through the same critique. Indeed, the problematic of disability in the Judaic, pre-Christian era, with which Stiker opens his account, already lends itself to just such a postconventional approach. In an exemplary openness with regard to the difficulty of what Foucault has named as genealogy, Stiker admits that it is not the social practices of the pre-Christian society as such that he is describing, but only those described in the Bible, which as he notes 'is a prodigious document but, at the same time, one that conceals' (1999: 23). But it does demonstrate the twist of *différance*, for as he notes of the contemporary texts, the disabled body is widely seen as both unclean – a status that may be held at a distance – and *polluting*, which implies the capacity to cross boundaries. In biblical terms, then, disability can be positioned as an abomination that is subject to an array of exclusionary *and* purification procedures. In the verses of *Leviticus* that set out the parameters of the Law of purity in detail, congenital malformation is particularly defiling, together with leprosy, surface tumours, scabs and other highly visible disturbances to the ideal closure of the skin (*Leviticus* 22).[8] Although, as he acknowledges, what was initially at issue in the Law may have been a prudent matter of hygienic protection, Stiker speculates that it nonetheless opens up a symbolic distinction between the pure and the impure in which certain disabilities fall on the side of the unclean and the ungodly (1999: 26). Malformations of the body, as the Old Testament makes clear, are as much a spiritual as material insult, and the sanctity of the Temple must be afforded especial protection:

> For whosoever man he be that hath a blemish, he shall not approach; a blind man, or a lame, or he that hath a flat nose, or anything superfluous... he shall not come nigh to offer the bread of his God.
>
> (*Leviticus* 21: 18–21)

Nonetheless, even in the chapters where most of the rules and prohibitions concerning defilement are laid down, there are warnings against mocking deaf people or putting obstacles in the way of those who are blind (*Leviticus* 19:14). Here, as elsewhere in the Greco-Roman world, where the exposure of *congenitally* malformed infants was sanctioned by law, some forms of sensory deprivation – perhaps those that are acquired – escape symbolic marking and do not figure as dangerously out of place.

A similar reading of the biblical texts, but one that is substantially more alert to the psychic underpinnings, is offered by Julia Kristeva (1982, 2000) who goes a step further in her attempt to recover what she calls an 'archaeology' of impurity. Like Stiker, she is well-versed in the anthropological function of taboo, which both scholars link to the socio-religious systems of ancient Judaic society. As Stiker puts it: 'taboo... relieves anxiety in the face of the unknown, the uncommon, the altered. A defense is needed; the remedy lies in purification' (1999: 26). The point, for Kristeva, of the excavation of the past is not to demonstrate the archaic otherness of early societies, but to trace the continuities of psychic and – to an extent – conscious, concerns. As she puts it, the 'so-called archaism trips us up more often than we think' (2000: 21). As Kristeva uniquely understands it, the whole nexus around the concept of impurity is overdetermined by the maternal, but she also explicitly recognises corporeal alteration – again with reference to Leviticus – as a category of abomination (1982: 93). She asserts that the significant concern is never with a perceived lack of health, however, but, as she makes clear in her study of the abject, with a potential failure of system and order. The problem in other words is not with some inherently impure substance or body, but rather with any potential breach of the boundaries of the clean and proper. The threat resides in contamination and pollution. Moreover, although material abomination might appear to be simply in an oppositional binary relation to the sanctity of the Temple and the logicality of the Law, the categories are interdependent: '[t]he one and the other are two aspects, semantic and logical, of an imposition of a *strategy of identity*' (1982: 94). Kristeva continues:

> the 'material' semes of the pure/impure opposition that mark out the biblical text are not metaphors of the divine prohibition... but are responses of symbolic Law, in the sphere of subjective economy and the genesis of speaking identity.
>
> (*ibid*)

What this implies in terms of disability is that there is an inherent and necessary complicity between 'whole' and disabled embodiment. Integration and exclusion are not simply mutually sustaining, but adhere in the same moment of coming into being.

In Stiker's account, the status of those with disabilities was radically transformed by the teachings of Christ. Although Stiker insists that there had always been an ethical obligation to counter-balance the religious taboos that conflated bodily anomaly with sin, the new Christian approach – both didactic and practical – neither excluded nor blamed.[9] Nonetheless, it is possible to discern the point of uncertainty about the degree and distinction of the otherness posed by corporeal difference re-emerging in the Christian Era, this time in the form of persistent philosophical discussions – again sanctioned by religious authority – about the 'natural' status of the aberrant body. It is not just that the varying morphology of disability and its range of severity posed the question of limits, but that even in the most radical cases of corporeal alterity, in which disability elided with the monstrous, the relation of that body to the natural order was by no means certain. Was it, on the one hand, a member of a category entirely outside nature, or simply a deviation, a freak, a *lusus dei* even, within? The debate is crucial for it has far-reaching implications concerning both the manifestation of the power of the godhead and the potential salvation of souls. A good part of the debate occupying the Church Fathers was centred on whether corporeal aberrance was an instance of god's power to vary the natural order at will, or alternately a break with that order that signified the exclusion of the fallen and sinful. To be absolutely different, in a different order of being, to be *other than* fully human, was to be beyond redemption – as were the sons of Ham, whose sin against god provoked a degeneration into animality (Friedman 1981). In contrast, the binary of natural/unnatural contained difference within the order of the same, and left open the possibility of salvation. On the basis of such a view, Augustine (1972), who taught that anomalous embodiment should exclude no-one from the community of the god-given, believed that at the last judgement and resurrection of the dead, the sick, disabled and deformed would be restored to the whole-bodiedness of normative morphology.[10] As the hallmark of European attitudes in the late middle ages and early modern period – a period rich in the archives of difference – otherness, then, was always an ambivalent, even undecidable, category.

Given the characteristic interest in all that appeared as either supra- or supernatural,[11] the nuances of scholastic debate concerning the status of the monstrous body spill over into more popular texts, such as the so-called monster or wonder books that circulated throughout Europe, particularly during the Renaissance and early modern period.[12] The temptation to view such texts as simply credulous, as the equivalent of *The National Enquirer* or *Sunday Sport*, is strong, but there is, I suggest, much more at work than the desire to shock and amaze, or even to entertain. There have been many unproblematised suggestions that prior to the normativities imposed by the modernist epistemology of embodiment, the existence of disability itself was not seen as oppositional. Such suggestions refer not to a reading of disability as a category that was absolutely other, beyond comprehension, and therefore removed from any oppositional relation, but to quite the opposite: the proposition that it was simply unremarkable. In that view, it is usually asserted that disability as such was just another taken-for-granted aspect of ordinary life that was conceptually – if not materially – assimilated to everyday expectations.[13] This may to an extent have been true of acquired disabilities in particular, and given that those meeting with disfiguring disease, trauma on the battlefield or accidental amputations, for example, could expect little by way of medical intervention to allay the effects, such morphological anomalies must have been relatively commonly encountered. Certainly modern-day western tourists coming across children with hydrocephalus or blind men and women begging on the streets of an underdeveloped nation may experience shock and amazement where local people do not; but the differential degree of familiarity tells us little about the level of psychic assimilation. In any case, even when difference is represented in a sizable minority, the response may be hostile, anxious or curious, or a combination of those reactions. Similarly, the visible presence of unmodified disabilities in Renaissance and early modern times did not forestall the commercial viability of the monster books that included wondrous images and accounts of a range of corporeal differences from the seemingly insignificant to most unusual.

As I understand it, any assessment of the veracity of such representations is irrelevant, for – whatever the stated intention – the primary appeal is not to a descriptive truth but more to the cultural imaginary of the writer or reader, including of course present-day scholars. The extant textual accounts themselves are highly conflicting and inconsistent even across similar forms of disability, and despite insistent editorial truth claims, there is little reference to what present-day readers would recognise as evidence, even in a weak form. The same illustrations, for example, are reproduced again and again to cover narratives that are wildly divergent in time, place, and content. Moreover, the question of valuation is itself uncertain for although congenital – if not acquired – disability was frequently classed as monstrous, an affront to the natural order of things, it is by no means clear that that was an entirely negative nomination. At the very least it inspired both fear and fascination. Certain forms of radical corporeal aberrance remained highly ambivalent in that they could be read in multiple ways that might on the one hand signal the presence of evil, but on the other, the exercise of god's extraordinary power (Shildrick 2002). As the mid-sixteenth-century works of the French surgeon, Ambroise Paré, make clear, the display of both mental and physical disorder might be taken to mark very human anomalies, often arising from the failure to observe the proper conduct of sexuality, but equally they could be a sign of divine intervention. In his famous list of the 13 causes of monstrosity, Paré (1982 [1573]) offers an explanatory system that already draws out several of the competing elements that surface in any history of disability. Nonetheless, it should be acknowledged that far from simply representing a series of existent superstitions, and for all his acceptance of the contingency of divine intervention, Paré himself is actively engaged in the search for a *rational* and systematised explanation of how certain bodily deformities might appear.[14] Though not yet a figure of modernity, he stands at the cusp of Enlightenment scientism. The grasp of knowledge and its attendant strategies of regulation are about to re-emerge in new forms.

The monstrous body, then, plays a key role in the construction and reinforcement of distinctions between normal and abnormal, and in the imposition of normativities. Although Foucault somewhat unexpectedly argues that disability is not coincident with the monstrous in that it has a place in civil or canon law (2003a: 64), it is clear from the texts of the early modern period that congenital disabilities at least were approached in the same epistemic frame as human/animal hybrids. Foucault's suggestion, moreover, that what defines the monster is a violation not only of natural law, but also of civil, canon, and religious law, recalls precisely the arguments that occupied the Scholastics with regard to morphological anomaly in general. It is not my claim that all disabilities were seen in the same way, but that they introduced the anxiety of an 'undecidability at the level of the law' (2003a: 65). On a Foucauldian reading of the breaks and fractures in history, the emergence of scientific and empirical rationalism – taken as characteristic of the post-Cartesian period – is closely associated with the relentless drive to categorise and establish clear boundaries of separation. As such, the positivist aims of modernist bioscience, which emphasise the elimination of ambiguity, the privileging of rational action, and, more particularly, the power to bring all bodies under a system of normativities have been highly influential in constructing and maintaining the distinctions between normal and pathological embodiment. At the heart of this supposedly new approach to bodies, in which the disabled body was a prime target, Foucault explicitly marks as significant the break between the traditional exclusion of the leper, left to his or her own devices, and the early modern inclusion of the plague victim, subjected to a range of normalisation processes:

> We pass from a technology of power that drives out, excludes, banishes, marginalises, and represses, to a fundamentally positive power that fashions, observes, knows, and multiplies itself on the basis of its own effects.
>
> (2003a: 48)

Yet, though the monstrous body is claimed by biomedicine in the interests of both repair and normalisation, and thereby loses much of its ambiguity, it remains, as Foucault notes of the nineteenth century problematic of abnormality, 'the magnifying model, the form of every possible little irregularity' (2003a: 56).

But is Foucault's own neatly boxed account to be trusted? Despite the widespread criticism that he misrepresents the historical 'facts', I would rather accept, from a position of postmodernist scepticism with regard to such convenient entities, that his outline is a highly insightful heuristic device. Nonetheless, despite the perspicuity of his exposure of binary thought around the concepts of normal and abnormal as fundamental to structures of postEnlightenment knowledge, I prefer to see that discourse as merely coming to dominance, rather than as a sudden irruption. What is of more significance, however, and what Foucault misses in his account of the 'inscribed surface of events', is the reason for the failure of the modernist episteme to fully encompass the problematic of disabled bodies. Although the mixture of curiosity and cultural unease given full public rein in the monster books may have become less authoritative in the light of rationalist and scientific discourse, those texts continued to circulate and to be reproduced in relatively unchanged form well into the eighteen century.[15] The motivating anxieties of such counter discourses speak to just that ambiguity that is figured by the anomalous body, a body that inherently resists reduction to either sameness or difference, to the natural or unnatural, and in extreme cases to human or non-human. Such undecidability is deeply unsettling to the cultural imaginary, particularly one that incorporates an image of the embodied self as whole, separate and invulnerable. The point is that concern with corporeal anomaly, and specifically disability in terms of my analysis here, is always more about the status of the supposedly ablebodied self than about the perceived corruption of the body or physical failings of the disabled person herself. Whether the response is individual or social, or indeed part of a dominant discourse like that of biomedicine, I would argue that it is always based on psychic anxiety. Like the monstrous, the disabled body is abject in the full Kristevan sense of the word. And regardless of what rationalisations might be offered in the attempt to hold categories apart – as, for example, in the major strategy of pathologising disability – the enforcement of normativities at the discursive level always falters because the binary system on which they rest is never fixed nor stable. In other words, the supposedly 'pure' categories of normal and abnormal, just as poststructuralism avers, constantly contaminate one another. The assault on the cultural imaginary is scarcely unique to the modernist era, however, but can be traced in all periods in which the ambiguity of the other's body threatens to spill over into the body of the self.

In illustration of my claim that the response to disability cannot be understood without reference to a transhistorical ontological anxiety operating at a psychic level, the discourse and practices of eugenics are particularly apt.[16] In post-Darwinian times, the notion of eugenics as the possibility of improving human stock by selectively breeding out anomalies initially caused few moral misgivings. On the contrary, the application of eugenic principles was seen not only as properly based in contemporary science, but as a utilitarian good in its own right. Nonetheless, what drove eugenic practice was always less the putative facts of bioscience than a series of both explicit and implicit value judgements in the realm of both the aesthetic and the moral. Moreover, although the terms of the debate have changed over time, and eugenics itself is widely discredited, contemporary 'biomedical' decisions are equally likely to be made on the basis of cultural values as to which disabilities are intolerable and should be eliminated at a genetic level or foetal stage, or which should be subjected to interventionary procedures. Though few authoritative figures today would openly advocate eugenics (or at least not using that vocabulary), it is clear that the deep discomfort provoked by any overt difference that breaks cultural normativities is still operative. What is at issue is that the category of disability represents what Paul Longmore calls 'delinquency in the practice of individual self-control' (1997: 153). In other words, it exposes the western fantasy of invulnerability and self-mastery, and threatens the illusion of the absolute sovereignty of the embodied self. The exercise of overt moral censure of those who are disabled is muted, but disability is still seen as an embodied incompetency that might overflow the parameters of its designated place. The movement between exclusion and assimilation that both Foucault and Stiker posit is always complicated insofar as neither operation is sealed against its other. Even within the most fully realised strategies of integration, such as inclusive education, workplaces, or theatre companies, the anxieties mobilised by the supposedly contaminatory potential of otherness cannot be allayed. Disability remains remarked, and thereby set aside.

With regard to the history of twentieth-century eugenics – and particularly in relation to the National Socialist programme – the bare outlines of the worst excesses of thought and practice are well-documented, and I shall recall them very briefly. The conflation of disabled, racial, and sexual others as both physically contaminatory and morally debased was the driving force in the social purity and eugenics movements, not simply in Germany but in many contemporary countries. Documentary figures are variable, but it appears that under the Nazis, over 350,000 of those classified as genetically inferior by reason of mental or physical disability were sterilised, with another 250,000 killed outright. The eugenic agenda predated the war years, but it was not until 1941 that internal public distaste finally halted the adult extermination

programme, although killings of children up to the age of 17 continued throughout the war (Gallagher 1990). The active part played by
the medical establishment in facilitating state policy both by select-
ing individuals and by carrying through the necessary procedures is
at once a shock to normative expectations of care, and a self-serving
social alibi for the lay population. The nexus of power/knowledge that
made such processes not only possible, but permissible, is predicated not
on a singular discourse, but on the micro-practices of power in which
every individual is implicated. Where social theory has been deployed in
an elucidatory capacity, one well-regarded explanation – associated initially with the Frankfurt School – suggests that the German state of the
1930s and 1940s represented a hyper-realised version of the modernist
project of instrumental rationality that seemed to justify, even mandate,
social engineering (Bauman 1989). The reference to supposedly inferior
human beings in brutally utilitarian terms as 'useless eaters'[17] supports
such a view, but fails to explain why the designation ever had epistemological currency. The internal justifications of the eugenic programme
clearly cannot be reduced to singular motivating causes, but it remains
necessary to ask why the citizens of even the most bureaucratic state
should assent to means that so clearly overrode the moral principles
of both Judeo-Christianity and the humanist Kantian ethics on which
European values are largely based. Certainly disabled people were dehu-
manised under the rubric *lebensunwertes leben* (life unworthy of living),
but are we not also able to hear in the term 'useless eaters' a hint of para-
sitism that arouses a psychic fear that, unless eradicated, the unwelcome
other will fatally sap the strength and vigour of the general population?
It seems pertinent to ask what psychic mechanisms are in play to create
such a sense of revulsion.

Given that large-scale eugenic programmes were a feature not just
of Nazi Germany, but of many other politically divergent western
European states – including such latter-day paragons of social welfare as
Sweden where a sterilisation programme continued until the 1980s – it is
clear that the argument from instrumental rationality is insufficient.[18] In
the United States too, between the years of 1920 and 1980, 33 states had
active laws allowing for the forcible sterilisation of not only 'the insane,
idiotic, imbecile, feeble-minded or epileptic', but often those with physical disabilities as well (Pfeiffer 1994; Mitchell and Snyder 2003). Such
programmes, moreover, which characteristically elided the deviance of
disability with criminality, were supported at the highest level, as the
notorious 1927 dicta of Supreme Court Justice Wendell Holmes makes
clear: 'It is better for all the world, if instead of waiting to execute

degenerate offspring for crime, or to let them starve for their imbecil-
ity, society can prevent those who are manifestly unfit from continuing
their kind' (cited in Block 2000: 246). And at the beginning of the
twenty-first century, bioethicists are still debating the rights and wrongs
of restricting the reproductive capacities of 'at risk' individuals deemed
incapable of making informed choices about sex, or of raising children who may result from it, by reason of their mental disabilities (see
Carlson 2001; Dowbiggin 2002; Gerodetti 2006). The cases tend to be
clustered around the issue of sterilisation, with healthcare professionals seeking a legal mandate to sterilise those – usually young women
or girls – whose mental competency is supposedly compromised.[19] The
rule of thumb adopted demands that evaluation be made in the best
interests of the named individual of concern, but, as with all recourse to
utilitarian decision-making, it is clear that the determinations, as well as
the precedents on which they are based, are as much about social expe-
diency as ethical concern. Occasionally such justifications are voiced, as
in the view of Peter Singer, Professor of Bioethics at Princeton University,
that there are good cost grounds for having fewer disabled people in
society: 'it does not seem quite wise to increase any further draining
of limited resources by increasing the number of children with impairments' (cited in Disabled Peoples International 2000, ¶3). Singer – who is
something of a bête noire for the disability community – is well known
as a utilitarian philosopher who delights in pushing the abstraction of
the harm/benefit calculus to its apparently rational extremes. What I
would suggest, however, is that there is something else at work here,
and that Singer's dicta exemplify not so much a high-minded instance
of rationalism as a rationalisation (Shildrick 2005a).

   It is precisely here in the midst of a public discourse ostensibly based
on reason that it becomes necessary to add in – at the level of both
the individual psyche and the cultural imaginary – that deep anxiety
pertaining to difference. And it is an anxiety less about an absolute separation of the categories of the normal and the abnormal, than about
the unbearable ambivalence of not being able to definitively settle on
difference. People identified as disabled provoke anxiety, not because of
their difference as such, but because they are too much like everyone
else; worse yet, anyone could become one of 'them'. In other words,
they defy the boundaries of sameness and difference and spread impurity through the normative categories. The dehumanisation of disabled
people – their reduction to 'useless eaters' – is, then, a denial of any
commonality with the normative majority that allows and implicitly
demands violent action against the threat of a disordering ambiguity.[20]

It is not my claim that such violence is always realised, but that the anxiety that drives it lurks as much beneath even seemingly beneficent reactions, such as a stated and most likely well-intentioned desire to improve the life of a disabled individual, as in overt eugenic intent. A growing range of reconstructive surgeries and other biomedical impositions purport to 'fix' the problem not just of radical anomalies such as intersexuality or conjoined twins (where there *may* be some clinical justification), but of a host of minor disabilities such as having a non-standard number of digits, or falling below normative developmental standards for infant growth. Such interventions speak to a desire, if not to eliminate, then at least to cover over embodied differences. As Rosemarie Garland-Thomson notes:

> these procedures benefit not the affected individuals, but rather they expunge the kinds of corporeal human variations that contradict the ideologies that the dominant order depends upon to anchor truths it insists are unequivocally encoded in bodies....[T]he medical commitment to health... has increasingly shifted toward an aggressive intent to fix, regulate, or eradicate ostensibly deviant bodies.
>
> (2002: 12–13)

That notion of either regulation or eradication (or less dramatically exclusion) surfaces clearly in Henri-Jacques Stiker's claims that by the late nineteenth century, people with mental illnesses were dealt with by 'exclusion and surveillance', while physically disabled people were subjected to 'regimes of recovery and assistance' (1999: 114).[21] His point is that for the latter category, the focus is on a return to societal and personal normativity, through technologies of rehabilitation, not least of which was the appearance of many new 'normalisation devices', such as neck braces, corsets, electrical stimulation, and prosthetic limbs. It is perhaps significant, although Stiker does not investigate it, that the drive to remake the body arose largely in response to the acquired disabilities and bodily devastation of warfare – first the American Civil War, and later the world wars of the twentieth century – rather than to aetiologically more disturbing congenital conditions.[22] For Stiker, in any case, the new emphasis figures not a humanitarian advance, but a more rigid application of societal constraints that act to efface difference by aggressively promoting normalisation, both informally, and through growing legislative, as well as biomedical, intervention. It is a strategy of sameness that promises greater equality to disabled people, yet only insofar as they are reintegrated into the norms and values of the dominant majority. As he puts it: 'Paradoxically, they are designated in order to be made to disappear, they are spoken in order to be silenced' (1999: 134), and further, '(s)pecificity and aberrancy are...forbidden and condemned' (136). The interventions, as Stiker himself notes without further analysis, are an effective way to evade the fear of strangeness, a fear that is no less apparent in contemporary society. Although the current focus is shifting decisively from a biomedical model of pathological embodiment that might be individually reclaimed to a social model in which contingent difference might be structurally accommodated, the outcome once more is that difference is erased. Both the *Americans with Disabilities Act* 1990 and the United Kingdom's *Disability Discrimination Act* 1996, which each respond in part to the growing ubiquity of the social model, attempt to efface the uncertainty of disability by promoting the view that everyone, including the disabled, can be productive workers. The construction of docile bodies, subject to control, standardisation, and predictability is accomplished as an act of social justice.

Despite its plethora of undeniable achievements, then, the SMD unwittingly perpetuates a historically situated pattern of disavowal. It is not simply that the model appears to devalue the phenomenology of embodied difference, such that issues of pain, desire, and affect are rarely incorporated, but that the emphasis is complicit in the denial of difference. Most importantly, in the pursuit of social and environmental adaptations, it covers over any recognition of the fragility and vulnerability, not only in the context of disability, but of all forms of embodiment. And it is precisely that desire to deny or disavow a vulnerability which refuses to remain hidden that generates anxiety in the individual psyche and cultural imaginary. The collapse of specifically different categories of disability into a generalised form of improper embodiment is conventionally set against the normative standard of the body perceived as closed, invulnerable, and separate. Yet that body, as psychoanalytic theory has long shown, is phantasmatic, at best an unstable image of integration, an ego-ideal, that belies what Lacan calls '*le corps morcélé*' (1977a: 4). Given its masculinist underpinnings, it is scarcely surprising that the normative model has been challenged by feminist theory in particular. The perception that women's bodies are always already leaky, and anomalous, threatening to overflow the proper boundaries of embodiment and separation, indicates not a gender-specific condition of being, but the insecurity of all bodies (Shildrick 1997). As Liz Grosz notes, female corporeality is characterised as 'a leaking, uncontrollable, seeping liquid: as formless flow...lacking not so much or simply the phallus but self-containment...a formlessness that engulfs all form, a disorder that threatens all order' (1994a: 203). And although the privileged order here may be that inaugurated in the modernist era, disorder, ambiguity, and uncertainty have always been productive of anxiety in western culture.

The failure of control apparently evidenced in disability, indeed the very undecidability of that embodiment, is, then, too close for comfort. Far from any acknowledgement of the intercorporeal constitution of all bodyliness that phenomenology proposes, or even the mutual engagement of existential status, disability must be disavowed, precisely *because* the disabled body cannot be wholly other. Against the desire to maintain self-control and to recognise vulnerability only as a quality of the other, disability at the very least reminds the majority of their own fragility. And it is not just in the context of unexpected illness or trauma, but as an inherent condition of being, a point Stiker takes up when he remarks that the disabled 'are the tear in our being that reveals its open-endedness, its incompleteness, its precariousness' (1999: 10). Nonetheless, a fuller analysis might go even further in figuring disability as a trace of the chaotic structure of early infancy which, though absent from consciousness, persists in the instinctive life of the adult, in what Lacan refers to as: '(t)his disarray, this fragmentedness, this fundamental discordance, this essential lack of adaptation, this anarchy, which *opens up every possibility of displacement*' (1988: 169, my emphasis). What is significant in such a structure is how it stands against the fulfilment of subjectivity. Although I shall not develop Lacan's formulation in relation to disability further in this chapter, it is, I believe, a potentially fruitful route for understanding the implicit violence that marks the encounter with the disabled body. In a discussion elsewhere, Lacan tellingly names aggressivity as 'a correlative tension of the narcissistic structure of the coming-into-being of the subject' (1977b: 22). In the hostility, evasion, and paradoxically in the fascination that greets disability, we might sense, then, a moment of self-recognition, or an intuition of the Lacanian mode of misrecognition that otherwise sustains the self in the symbolic. It is as though each one knows, but cannot acknowledge, that the disabled other is a difference within, rather than external to, the self.

That ambivalence, and its associated anxieties, surely marks a recurrent pattern manifest in discrete historical moments that, though shaped by highly variant contingencies, all bristle with unease. If the desire to construct distinction and separation must inevitably fail, then those too-familiar others who are disabled will be invested with all the anxieties and fantasies that operate at a socio-cultural, interpersonal, and ultimately an intra-psychic level. As such, disability is a highly complex and intrinsically ambiguous designation, that cannot be addressed adequately by positivist and binary-based models of analyses alone, whether those of biomedicine, historical narrative, or social construction. Whilst having no pretensions to articulate an ultimate truth, any genealogy of the discursive constitution of disability must engage it as the site of multiple cultural and psychic investments. In signalling the usefulness of a broadly psychoanalytic approach, then, it must not be forgotten that genealogy seeks to uncover the domain of power/knowledge as epistemically layered, and perspectival. There is no singular explanation, and no certainty in the quest for understanding, but only the dis/abled body in all its contingency and undecidability. In keeping with my interdisciplinary approach, I shall return to Lacan in Chapter 4 where psychoanalysis is brought to bear on the specific question of disability and sexuality, but for the next chapter, which introduces that vexed conjunction, the focus is on the disciplinary strategies and governmentality directed against it.

3
Contested Pleasures and Governmentality

In the contemporary western world, considerations of sexual pleasures and sexual desire in the lives of disabled people play very little part in lay consciousness, and practically none in the socio-political economy. The most significant exception is when the negative reading of such concerns serves to activate an inclination to contain and control supposedly troublesome expressions of sexuality. The problem is that in the context of mainstream values, the conjunction of disability and sexuality troubles the parameters of the social and legal policy that purports both to protect the rights and interests of individuals, and to promote the good of the socio-political order. In addition to its theoretical framing, this chapter illustrates that problem with reference to some very substantive issues, and while its focus is firmly on the Anglo-American context, I suspect the argument could be extended to other locations where heteronormativity is the dominant force. The concern of both social policy and law is to encompass the bodies of all within a governmental grasp, yet clearly some forms of corporeality exceed the limits of what is thinkable. It is, as I have suggested, as though the very being of anomalous embodiment mobilises both an overt and an unspoken anxiety, an anxiety that is at its most acute in relation to sexuality. The outcome is the strange paradox evident in western society that alternates between denying that sexual pleasure has any place in the lives of disabled people, and fetishising it. Both responses constitute a refusal of sexuality as a regular element in disability experience, an effective silencing that damages not just self-esteem, but – bearing in mind Merleau-Ponty's evaluation of sexuality as a modality of existence – all aspects of the capacity for self-becoming.

In its mainstream form, social policy signals a no-nonsense pragmatism that works together with law to authorise a discourse that is wholly

normative in the sense of being both prescriptive and boundary defining. And although that intertwined discourse may reasonably claim to
encompass simple differences that are able to operate within the same
overall structure of value and judgement as the standard form, it struggles and ultimately fails to respond to marginal figures who transgress –
either by design or by necessity – the parameters of heteronormative
society. What is at stake is an ethical matter,[1] but it is not one that can be
satisfied by a facile appeal to equality, for that move must always be compromised by its implicit acceptance of a socio-normativity that is both
reliant on and hostile to difference. Rather, the responsibility is to promote the good of corporeal diversity. In talking about *disability* as such,
we should be cautious, however, that a term which bundles together
a large variety of disabling conditions into a single category runs the
risk of erasing the specificity of how different disabilities may impact on
sexuality. Nonetheless, part of the task is to uncover precisely the universalised attitudes and values, the taken-for-granted assumptions, and
the likely negative modes of representation that lie behind mainstream
perceptions of the so-called facts regarding disability and sexuality. The
issue is not simply to describe the operation of particular attitudes that
are in circulation, but to always question why meanings around disability and sexuality are constructed as they are. Indeed, insisting on
positive representations of disabled people rather than ones that devalue
their subjects uncovers a very different understanding of the body that
opens up the parameters of sexuality for everyone, regardless of individual embodiment. The implications for social policy are considerable
and challenge both its utilitarian approach and the principle of equality,
which together have grounded a socio-political economy of disability
predicated on rehabilitation or compensation. Aside from the concerns
raised in the preceding chapter about the shortcomings of any assimila-
tive discourse, in addition those goals leave no scope for considerations
of affective differences in realms such as sexual expression. I argue in
contradistinction that only a response that recognises and enables the
full embodiment of disability will open up the problematic to the issue
of sexuality.

My approach to the challenge draws on strands of cultural theory that
may sit uneasily with the concerns of mainstream social policy, but the
point is that the arena of disability and sexuality is so constrained at
present, both in theory and in practice, that all critical resources must
be brought to bear to effect any radical changes. One of the most useful
theoretical perspectives that can be deployed to throw light on the motivating forces of social structures is that associated with Michel Foucault,

who stripped away the illusion of fair and measured decision making to expose the relentless *governmentality* at the heart of the social order. In speaking of governmentality, Foucault (1979b, 1988b) moved beyond the image of the state as a coherent apparatus of juridical power to address the multiple ways in which social relations are ordered and controlled through a network of institutional and personally directed regulation. At the same time, he allowed that certain forms of self-governance can resist the power of governmentality, and although he did not address contemporary disability directly, Foucault's approach is proving fruitful for theorists who seek to go beyond the traditional paradigms of disability studies.[2] As before, however, I shall argue that the move does not go far enough. Specifically, I seek to examine what lies behind conventional responses to the problematic of disability and sexuality by engaging with a cultural imaginary that privileges corporeal wholeness and predictability above any form of bodily anomaly, and that supports fears that non-normative sexuality in any form is always a potential point of breakdown in a well-ordered society. While most of us would accept that, as individuals, we are driven at least in part by subconscious and psychic impulses, it is less commonly recognised that institutional practice too is subject to unacknowledged, or even unknowable, influences that may result in practices and policies that have little rational justification. The power of the cultural imaginary – which Rosi Braidotti defines as, 'a system of representation by which a subject gets captured or captivated by a ruling social and cultural formation: legal addictions to certain identities, images and terminologies' (2006: 85–6) – is a potent force that can no more be ignored in social policy than in literature or film. It is in part for this reason that despite my discussion of the various options in progressive social policy that open up the issue of sexuality, I am by no means convinced, as will become clearer, that the so-called solution to the invalidation of disability and sexuality is to be found in the extension of socio-political recognition and regulation. The call for sexual citizenship that many activists now make on behalf of the disability community is an inherently ambivalent demand that even as it aims for new freedoms and choices must inevitably invite new forms of governmentality. From a Foucauldian perspective, it could not be otherwise, and it is crucial always to consider the dangers potentially interwoven with the projected gains.

The point is that sexuality in any case is a major site of state intervention and control, with a range of socio-political instruments regulating its permissible forms and location, its public representation, the nature and validity of individual consent, its participation in an exchange economy, its validation in institutional structures, and so on. As Foucault comments on the eighteenth-century emergence of such processes: 'Sex was not something one simply judged; it was a thing one administered. It was in the nature of a public potential; it called for management procedures' (1979a: 24). The category of sex, then, is always politically invested, but more than that, as Foucault makes clear in *The History of Sexuality* (1979a), it is constructed by hegemonic power in the apparently natural form of heteronormative practice and desire. At the same time, however, there is always something excessive about sexuality that resists categorisation and containment. And it may be precisely because of the anxiety surrounding the domain of sexuality – an anxiety that relates to the inherent risk of losing self-control and self-definition in the intimate embrace of another – that it is so highly disciplined and regulated, so shot through with performative constraints. In short, despite its frequent designation as a private matter in which the state should have little or no interest, sexuality has long been at the heart of governmentality, caught up in all those mechanisms at work in the construction and maintenance of the sociocultural order. And yet, sexuality is a point of official concern not so much for what it does as for what it is. It is the very corporeality of sexuality that provokes concern. As such it would be no surprise to find that the sexuality of people experiencing disability – people, that is, whose very embodiment already strains at the limits of the normative – should be perhaps even more closely scrutinised. As Henri-Jacques Stiker puts it: 'This cannot be a private matter.... The differing body is socialised' (1999: 40).

The paradox, then, is that social policy and law alike are surprisingly silent with regard to the specificity of sexuality in the context of disabled bodies. The productive nature of power – which, as Foucault (1979a) asserted (using homosexuality as an exemplar), names and promotes new categories the better to control them – seems to have passed by without effect. But could that very absence of intervention have meaning? As Foucault put it:

Silence itself – the things one declines to say or is forbidden to name, the discretion that is required between different speakers – is less the absolute limit of discourse ... than ... an integral part of the strategies that underlie and permeate discourses. (27)

Any notable silence, then, must surely convey a negative meaning, implying in the case of disability that any reference to a specific sexuality is missing because disabled people are being actively constructed as non-sexual. Such silence is, in effect, management by non-recognition. The paradox of that absence becomes clearer when one considers that in other areas – such as employment, access to goods and services, education, and so on – the nexus of disability gives rise to much explicit and dedicated legislation and policy, yet the major, but limited, concern in the area of sexuality and sexual relations appears to focus primarily on regulating and monitoring the reproductive capacities of men and women with developmental disabilities. Where that monitoring results in potentially severe restrictions that may deeply undermine the enshrinement of reproduction as a fundamental right of humanity, it demonstrates state power in a more classically repressive form that is at least open to challenge. In contrast, an insidious management by silence is extremely hard to contest. Even in the authorisation of sterilisation, for example, the juridical language betrays a certain inability to encompass disabled people engaging in sexual relations at all.[3] What both modes of repression demonstrate in those in the mainstream is at very least an ambivalence, or more clearly a disinclination, towards recognising and giving value to difference at the margins. Where the influence of a cultural imaginary is at work, any disruption caused by an insult to its comfortable order may be negotiated by the interlocked strategies of both explicit management and silencing. I would even go so far as to contend that it is the issue of silence in the mode of psychic disavowal that in turn partially drives governmentality.

Let me first, however, make clearer the manner in which legal and social policy is fully implicated in the array of disciplinary and regulatory techniques directed towards the sexual pleasures of the individual body (Foucault 1979a, 1980). There is, as Foucault suggests, always the possibility that 'bodies and pleasures' (1979a: 157) might subvert normative stability. Indeed, the two strands are necessarily mutually generative. The coming together of disability and sexuality, then, may signal a problem to be at least managed if not silenced altogether, but at the same time it remains uncontainable precisely because sexuality in any form is always in a dynamic state of process that is neither predetermined nor fully open to performative intent: no-one acts entirely as she or he chooses. Instead, as Foucault argues, sexuality is 'organized by power in its grip on bodies and their materiality, their forces, energies, sensations and pleasures' (1979a: 155). In other words, even as power acts to cover over anomalous forms of sexuality, it serves to generate

a resistance that transforms the discourse in general. I am not suggesting that any of these outcomes constitute a conscious strategy on either
front; indeed the very fact that governmentality and lack of recognition appear to pull in opposing directions underscores the psychic forces
in play. As a result, the problem that emerges for disability activism is
that the goal of forcing a reluctant sociality to acknowledge the actual
needs and desires of disabled people has no clear-cut trajectory insofar
as the very moment of speaking out risks generating a new category subject to discipline and regulation. I will turn to this tension shortly, but
first want to explore how both the non-disabled majority and disabled
people themselves understand and experience the nexus of sexuality,
pleasure, and desire.

As far as the state is concerned, the quasi-disavowal of sexuality in the
context of disability plays a part in the maintenance of the normative
attitudes that shore up the supposed stability of a social order founded
on heterosexuality and nuclear family life. For western societies in particular, the accepted standard for sexuality is a monogamous relation
between two adults of opposite genders whose sexual practice is conducted in private and is primarily genitally and reproductively based.
In addition to those like lesbians and gays who choose to resist socio-
cultural norms and have traditionally been perceived as a threat to the
very fabric of society, there are others – people who literally cannot fulfil
normative expectations by reason of their embodied difference[4] – who
are positioned as sexual outsiders. And although the experience of variable disabilities can have anything from a profound physical affect on
sexuality to virtually none at all, sexual normativities – at least in authorised sexual practice – are set up in such a way that the particularity of
each person is largely over-ridden, with the result that the majority of
disabled people are positioned as sexless. Despite all the evidence to
the contrary, one widely held perception that persists is that to have a
severe disability precludes any form of functional sex, while sexual pleasure and desire are simply beyond consideration. Several studies have
indicated that the non-disabled majority – and they may include professionals in the field of social and legal policy as much as lay people[5] –
quite literally believe that men with spinal cord injury, for example, can
never achieve erection, that women with congenital skeletal disorders
do not experience orgasm, or that cerebral palsy necessarily precludes
giving birth. What is meant by functional sex, of course, concerns
mainly reproductive intent, but even that limited designation simply
disregards the fact that many disabled people with extensive impairments are both fully capable of, and committed to, having children and

families of their own. Regardless of less rigid attitudes with regard to sexuality more generally, when it comes to disabled women in particular, powerful and entrenched social values that privilege reproductive sex re-emerge to cast those women's so-called tragedy as the assumption that they cannot become mothers. In reality, however, what is at issue is not so much that the outcome of sex does not follow a normative path, as that, for disabled people, sexual expression itself – the mode of sex – must often take a differential form.

It is scarcely surprising that the general negativity of public representations of disability and sexuality – which in a classic binary structure both construct and maintain a powerful series of social norms – should simply seem to reflect the facts of life for disabled people. Despite increasing portrayals of disability in the media in the interest of diversity, relatively few of these provide positive representations of disabled people in a sexual context.[6] Aside from some rare exceptions, such depictions more often invoke an implicit pity stemming from the notion that disability signals that sexuality has either been lost or never experienced. But it is not a failure of perception and imagination that exists only for the majority; not surprisingly, many disabled people, growing up in a societal context that silences, devalues, and distorts their sexuality, have internalised many aspects of that negativity. Where children categorised having a disability may be excused from sex-education classes; where disabled adolescents may find their social encounters closely monitored by both professionals and parents, ostensibly to protect them; where sexual healthcare facilities have proved unwelcoming or even inaccessible; where disabled adults are infantilised and deemed incapable of making their own sexual choices; and where the sexually active disabled person is labelled as shameful or disgusting – in such a society, the reluctance of those most affected to become involved in sexual politics is not difficult to explain (Shildrick 2004a; Parritt 2005).[7] Nonetheless, alongside the public absence of discussion of sexuality, there has flowed another, more hidden current in which disabled people themselves have talked about their own sexuality at conferences, in workshops, in online discussion groups, via artworks, and occasionally in journal articles. Publication of the book *The Sexual Politics of Disability* (Shakespeare, Gillespie-Sells and Davies) in 1996 marked perhaps the first substantial opening up of the field of sexuality and disability into a politicised sexuality that had previously received only sporadic attention. Since then, what has begun to emerge in response to a general liberalisation around sexuality is a theoretical change of direction in disability studies that moves away from the burden of guilt and shame that the medical model has implicitly endorsed to embrace the non-normative, and to generate the very different reality evidenced by positive self-representations of disabled people. At the same time, specific empirical studies (Nosek *et al* 2001; Shuttleworth 2002) have uncovered a level of sexual expression that belies the myth of asexuality.

It would be easy to assume that the widespread *ignorance* of both the sexual actuality and potential of disabled people that persists despite such moves was no more than that, but in a socio-cultural context that has trouble acknowledging irreducible differences of all kinds, the benefits of a better education about how non-normative others could or do express their sexuality seem limited. None of us is able to escape the coils and effect of normative discourse, and it takes courage to resist dominant attitudes and values and insist that things could be otherwise. While loving and caring relationships involving disability on one or both sides are usually – though not always – approved and recognised, and functional sex may be reluctantly admitted by the non-disabled majority and increasingly claimed by those who are disabled, the real problem lies in the widespread disbelief or denial that sexual pleasure could be any part of such relationships, still less that desire could be valued for its own sake. It is where the sexual intention concerns not the goal of reproduction, but the expression of desire as such, that the disqualification of disabled people from the discourses of pleasure is at its most acute – an absence that can be as marked in the disability activist agenda as in mainstream concerns.[8] I shall return to this, but first it is worth exploring in more detail both the advances and the constraints that have shaped the recent sexual politics of disability. In particular the emergence and subsequent dominance of the social model of disability has been the occasion for some very impressive improvements to the conditions of everyday life, whilst at the same time serving to discourage exploration of some very specific concerns in which sexuality in all its aspects – relationship, identity, expression – is a major factor.

In broad terms, the theoretical and activist challenge to the medical model of a pathologised body as the meaning of disability has successfully replaced that image with a model that insists on consideration of the injurious effects associated with the social construction of a devalued identity. As I have already outlined, the drawback of the shift to a recognition of the contingent obstacles and barriers that society places in the way of those who cannot function according to normative standards has been that the actual significance of differential embodiment – a significance never captured by the medical model – may be easily overlooked. More specifically, the move away from a focus on the pathologised body has often sidelined the body as such and played into the existing tendency of disabled people to discount the need for recognition of their sexual becoming. In recent years, the shortcomings of the social model of disability have been more widely discussed (Shakespeare 2000) with a greater emphasis given to understanding the phenomenology of embodiment (Paterson and Hughes 1999; Price and Shildrick 2001; Snyder and Mitchell 2001). In short, as I discussed in some detail in Chapter 1, bodies matter not because we live in them, but because the experiences that constitute the self are always embodied. As the sexual activist and theorist Eli Clare comments, 'in defining the external, collective, material nature of social injustice as separate from the body, we have sometimes ended up sidelining the profound relationships that connect our bodies with who we are and how we experience oppression' (2001: 359). The more particular problem, then, is that the social model analysis can neither fully account for, nor mobilise concern for, the silencing and invisibilising of the sexuality of disabled people. What is at stake clearly encompasses more than the removal of obstacles, and alternative approaches that emphasise that living in the world is always a matter of embodiment are better positioned to open up and address the domain of sexuality as part of the wider problematic. Nonetheless, given my own further interests in the operation of a cultural imaginary, I think that although phenomenology is essential, it is not sufficient by itself. What is required is an intersectional analysis that considers the interlinked significance of various perspectives, as well as the places in which a particular analytic model takes precedence.

For many disability activists the slow moves over recent years towards what is called independent living – which signals a mode of existence that is structured by personal choice and the self-administration of welfare benefits – has delivered many improvements in the way in which disabled people are able to manage their own lives, not least in the area of sexuality. Instead of being subject to the supposedly best-practice standards imposed on them by external agencies – be they state controlled or commercial – recipients of the UK *independent living allowance*[9] are now theoretically able to set their own priorities, which may well include allocating resources for the satisfaction of desires as well as needs.[10] Clearly, such provisions open up extensive new arenas to the search for personal fulfilment and, more importantly, they break with one of the major obstacles to disabled people being enabled to engage fully with their own sexuality. In a society self-consciously committed to equity, the conceptual split between what is classed as a *need* and what is classed as a *desire* plays into a distinction between those claims or requirements that should be addressed versus those that may be reasonably set aside. The convention in western societies is that publicly funded benefits of whatever sort – be it education, health care, or welfare payments – should be directed primarily towards the meeting of needs, with mere wishes or desires confined largely to the realm of private provision. Other than what is deemed to be a legitimate state interest in and therefore support of heterosexual reproduction, matters relating to the sexual satisfaction of the individual have had little or no purchase on public funds. More specifically, the sexual lives of disabled people have not been considered a significant part of the social policy responsibility.[11] The fundamental question of whether a distinction between needs and desires can be justified in any case is scarcely addressed, and the manifest inequity of effectively denying to disabled people the experiences that the normative group may take for granted has been discounted. As a consequence, there is at an ideological level a ready explanation for the state's disinterest in making interventions that either compensate or accommodate for such matters.

For all that, the recent emergence – at the level of social policy – of a newly positive response to the notion of independent living marks a significant change in the hitherto repressive attitude as to what constitutes a full and satisfying life for a person with a disability. Nonetheless, although it is theoretically clear that self-managed funds allow for the purchase of sexual services, for example, as one possibility among others, that development is one which may trouble not only the cultural imaginary, but also the law. Certainly there is a socio-cultural anxiety at issue here, but also a social policy and legislative context that directly, albeit unintentionally, disallows and potentially criminalises the behaviour of at least a minority of those living as disabled and their personal assistants. Several issues of concern arise here in which the general provisions of UK law regarding sexuality may impact differentially on disabled people.[12] It is not that any relevant cases have come to court as far as I can ascertain, or have even been treated as a police matter, but that where any individual appears to stand outside the law, his or her behavior will be regulated regardless of substantive sanctions imposed. And although I say his or her here, it is predominantly masculine sexuality that is at stake. At the lowest level, a case in point would be the difficulty that many disabled men in care facilities have in gaining access to pornography. Many anecdotal accounts attest to the fact that caregivers frequently refuse to facilitate access, citing reasons of personal morality, albeit backed up by an instrumental power that ensures that their own views, rather than those of their clients, will prevail. And whether or not it is directly invoked, the law on possession and supply may well sustain their prohibitions. In a similar way, if a disabled person wishes to use the services of a sex worker and is unable to negotiate the transaction alone, either by reason of restricted mobility or communication skills, then any caregiver or assistant who facilitates the contact may be acting on the edge of legality. Now, quite clearly, many present-day managers and staff in group homes have no compunction in setting up such contacts if requested to do so, and could no doubt provide good ethical reasons for their actions, but nonetheless they may well be in theoretical breach of the law against sexual offences.[13] As such, there is a ready justification for anyone who wishes to impose restrictions. While the motivation for refusal to comply with client wishes is unlikely to be fear of actual legal proceedings, it is not self-evident that the objection is simply to pornography or the operations of prostitution per se.[14] Given the cultural anxiety around disability and sexuality, the restrictions, I suggest, are just as likely to be motivated by some degree of particularised disgust.

One major area of social unease – and it is one that has particular relevance to the notion of independent living – is undoubtedly that of facilitated sex. Once the disabled person becomes a direct employer rather than the passive recipient of services paid for by a third party, he or she is in a far stronger position to hire only those assistants who seem comfortable with sexual issues, and who are aware that they may be asked to provide sexual assistance.[15] That category might cover the fairly routine tasks of helping with make-up, undressing the employer for a sexual encounter, or bathing him or her after sex. Some disability activists go further, however, to argue that assistance should include things such as giving help with masturbation if requested by someone unable to achieve it alone; accompanying the disabled employer to sex clubs; negotiating prices and services with sex workers; and actively supporting him or her in the position in which they can best achieve sexual satisfaction with a partner. The very notion of facilitated sex is highly controversial not only outside, but also within the disability community, which inevitably is caught up in exactly the same processes of power-knowledge about what constitutes acceptable sex as the wider public. What is at issue here, I would suggest, is that while socio-cultural mores and the law broadly support a normative image of sexuality as heterosexual, private, ideally reproductive, and above all autonomous, facilitated sex – by definition – cannot be wholly private or self-directed. It all too clearly draws attention to the difference of anomalous bodies. One not unexpected consequence of this is that disabled people who avail themselves of sexual facilitators may feel constrained to strive for an appearance of the very independence and self-determinacy that mark the modernist values that oppress them. As Hughes *et al* note, the transformation of care relationships, under the rubric of independent living, into contractually based personal assistance buys into 'the logocentric and patriarchal heritage of the enlightenment. It might also mean that the ethical imperative of recognition of the other... is left out of the moral equation' (2005: 268). The disabled employer feels herself at liberty to behave as though her sexual encounters were indeed private and self-directed, thus effacing any sense that the phenomenological intercorporeality of the event must include the '3rd party' presence of the facilitator. The problem is not of course limited to assistance in sexual scenarios, but given that the public discussion of sex practices and, more particularly, of sexual variation is still a strong taboo in many western societies, then the very notion of such hands-on involvement is particularly disturbing for both the wider public and the participants.[16]

In one small qualitative study (Earle 1999) that focused on disabled university students and their personal assistants – and it is one of the very few research papers that has explored facilitated sex in a non-nursing environment – it is made clear that the task of the assistants was to provide personal, social, and academic support. The students, all except one of whom were male, not surprisingly saw that task as encompassing sexual exploration – an area that had been restricted or denied to many of them in the past. As Earle noted, however:

> personal assistants may refuse or be reluctant to facilitate sex, when they believe the student's sexuality is morally reprehensible... (most likely) when the sexuality of disabled people deviates from the generally accepted norms of conventional heterosexuality and sexual restraint.
>
> (1999: 312)

The difficulty, of course, is that even those disabled people who are sexually conservative in their own identities may find that their personal options for sexual expression may by necessity fall outside normative practice, not least in requiring active assistance. While some participants in Earle's study – who all occupied a role that represented the progressive side of social policy – were willing to facilitate requests, many were confused, defensive, offended, and dismissive about their employers' needs for sexual assistance. One of the most resistant helpers, a young man named Derek, was noteworthy for his argumentative invocation of the distinction between sexual desires and physical needs. He insisted that because the two are not comparable, he had effectively no responsibility towards meeting his client's 'wants'. As he explained:

> I mean if you don't get your pee or your shit out you're going to explode. You know you're going to be in serious discomfort you're going to be sitting in a mess for the rest of the night, whereas, you know masturbation...

(Earle 1999: 316)

Even though independent living can encompass sexual interests, then, the success of the contractual arrangements with personal assistants relies heavily on the pre-existing conscious attitudes – not least those surrounding non-normative sex – that in turn speak to a somewhat less accessible cultural imaginary. And it would not be impossible to see in the discomfort of the assistants a glimpse of an abject affect, a sense of disgust that marks the apprehension of the 'improper' body.

The issue of facilitated sex also raises the more readily acknowledged question of how far it is justifiable to exclude certain people from the expression of their sexuality when the situation could be easily alleviated. The point made by activists is that disability should be no bar to sexual citizenship, which is positioned as a right on the same level as political citizenship. Tobin Siebers, for example, argues that the recognition of such a right would not only 'advance the cause of other sexually oppressed groups' (2008: 136), but instantiate a more secure position in society. As he puts it: '[disabled people's] status as a sexual minority requires the protection of citizenship rights' (2008:154). Currently, however, social policy has largely exempted itself from such problematic questions by failing to engage directly with the sexuality of disabled people at all, and most arrangements involving facilitated sex are made privately. Even so, the fear of sexual abuse can occur on either side, both in traditional care facilities and in the more private environment of the disabled person's own home. Several studies (Razack 1994; Beail and Warden 1995; Curry, Hassouneh-Phillips and Johnston-Silverberg 2001) have documented the widespread prevalence of sexual assaults on people identified as disabled, but the issues arising from the explicit negotiation of sexual assistance are much more likely to arouse anxiety about personal, ethical, and professional boundaries. The relationship of power between the two parties is by no means straightforward, particularly when the one who is the employer is also the one requiring services that implicitly contest the socio-cultural normativities of sexual exchange. Moreover, it is not insignificant that care work in general has a low status, and is often not only poorly regulated but also unsupportive of employees. Care workers themselves may well feel exploited by a system in which they have low pay and over which they have little control. That both state and private agencies typically attract assistants who are poorly educated, or from ethnic minorities whose sexual mores may differ from those of their clients, serves to exacerbate what is in any case a point of tension. Given this situation, little incentive exists to negotiate the question of unmet sexual desire.

I have referred already to some of the direct legal problems that may arise around disability and sexuality – problems that can be exacerbated in the case of facilitated sex if the disabled employer is a gay man. Although consensual acts of homosexuality are no longer always a crime in many western jurisdictions, a homosexual act remains illegal under the *Sexual Offences (Amendment) Act 2000* in the United Kingdom, for example, if it takes place in a situation deemed not to be private. Given that the presence of any third person or persons is understood to break that privacy condition, then clearly gay disabled sex is, strictly speaking, illegal if it is facilitated by a personal assistant whose physical presence is required. Perhaps because of their traditionally restrictive mores and laws regarding sexual expression, the United Kingdom and the United States are perceived as lagging behind other western countries on the issue of facilitated sex: it is worth noting that not only in the Netherlands (Valios 2001) but also in Denmark and parts of Australia, the use of trained sex workers and sexual surrogates for disabled people is sometimes subsidised by the state. Even so, provision is patchy, and as a recent article by Jan Browne and Sarah Russell[17] on the situation in Australia makes clear, both non-specialist *professional carers*[18] and their clients experience significant reluctance to cross certain normative sexual boundaries. In their brief consideration of assisted masturbation practices, Browne and Russell report: 'It was not only carers who were "too embarrassed". Participants described being "too shy to ask"' (2005: 381). The research found that both parties were anxious about crossing the line that supposedly marked the difference between the professional engagement of an assistant with a disabled client and a personal, and potentially sexual, relationship. In policy terms, caregivers were further constrained by organisational guidelines that discouraged them from assisting their clients with sexual needs, including, it appears, helping them to access outside commercial sex services (Browne and Russell 2005). Although they do not specify the geographical limits, Browne and Russell's study seems to have centred on Melbourne, so it is difficult to know whether research in other state jurisdictions in Australia would yield similar results. Certainly Australian law at the federal level is significantly more liberal with regard to differences in sexual practices than UK or US law, but the removal of the threat of criminalisation does not necessarily remove issues of professional misconduct.

Insofar as they rely on the implementation of new policies at the organisational and state level, the paradigms of independent living and facilitated sex might seem an unproblematic advance for which all disabled people should be willing to struggle. Nonetheless, the naming of such changes in public policy as progress – even though they may indeed improve the immediate, material conditions of disability – is to overlook the considerable risks of attaining recognition for sexual interests at the cost of a certain normalisation. What I am suggesting is that even the most seemingly benign developments arising from policies of liberalisation with regard to sexual matters may merely mask a persistent and underlying failure to make space for that which resists the closure of final classification. As Stiker (1999) makes clear in his survey of the move towards the rehabilitation of disabled people in twentieth-century France, that new emphasis figured not so much a humanitarian advance, as a more rigid application of societal constraints. By aggressively promoting normalisation, through interventionist strategies in the legislative and policy field, the difference that was irreducible was effectively effaced. In other words, strategies that promised greater equality to disabled people relied on imposing a mode of sameness that allowed only for assimilation and integration. As Stiker sardonically notes: 'Paradoxically, they [disabled people] are designated in order to be made to disappear, they are spoken in order to be silenced' (1999: 134). The issues that Stiker raises both resonate with emergent tensions within contemporary disability politics with regard to the need to raise the issue of sex and sexuality as a public matter, and reflect recent similar struggles that have occurred around lesbian and gay demands for sexual citizenship. Bearing in mind Foucault's dictum (1979b) that sex always incites management procedures, any raising of the prospect that sexual citizenship might extend specifically to disabled people (Shakespeare 2000; Richardson 2004; Sherry 2004; Siebers 2008) generates an urgent need to discuss whether that putative gain would be to further invite the influence of governmentality.

The concept of citizenship in terms of western democracies is always to some degree a disputed one, and the further notion of sexual citizenship is by no means self-evident (Berlant 1997; Weeks 1998; Phelan 2001). For scholars concerned primarily with non-normative sexuality, the latter may mean no more than the full integration of lesbians, gay men, and others living outside heteronormativity into a shared *civic* identity that in itself does not specifically address sexuality. Such an absence reflects to some extent the traditional understanding of citizenship as an issue of public recognition with all its attendant rights and duties, whereas the intimate matters of embodied identity are assigned to the private realm of personal choice. Chris Bacchi and Chris Beasley (2002) go further in asserting that the state deliberatively renounces intrusive control over the citizen's body, as a measure of his or her political autonomy. It is only when the political subject is deemed incapable of such control that regulation is thought to be justifiable. As they put it: 'The "control over body" subject is equated with "citizen", whereas those reduced to their bodies are constituted as lesser citizens' (2002: 326). As such, it is easy to see that disabled people – who are habitually marked as having some kind of bodily vulnerability and dependency – are among those who would be subjected most readily to external strategies of body management, including, inevitably, disciplinary measures directed towards their sexuality. But though I find Bacchi and Beasley's analysis appealing, I would dispute the clear distinction they make between the controlling and the controlled subject. As I understand it from a more committed Foucauldian perspective, no-one, whatever their form of embodiment, escapes the web of regulatory power that is directed towards all aspects of bodily identity, comportment, and behaviour. The distinction between public and private that appears to exempt the personal aspects of sex from the disciplinary gaze of legal and social policy is an illusion. To have a disability may be to invite more extensive surveillance, but it is a difference in degree, not in kind. When Bacchi and Beasley claim, therefore, that 'embodied citizenship... appears to be a contradiction in terms' (2002: 326), I would counter that on the contrary citizenship already encompasses or, rather, lays hold of bodies, so long as they fit within normative standards. The consequence is that for non-normative others to make a claim to citizenship is not a route to securing privacy in or personal control over bodily matters.

Despite that area of disagreement, Bacchi and Beasley do, however, indirectly highlight the problem that many other advocates of sexual citizenship have overlooked: as they see it, the push for that status seems to be more about gaining legal and social rights for a sexual *identity* than for the protection of specifically non-normative sexual *behaviours*. As such, the appeal to sexual citizenship can be seen as retrogressive, and critics of the approach tend to align themselves with queer politics.

It is a move that is proving an increasingly productive alliance for radical disability scholars (see McRuer 2003, 2006; Sherry 2004; Shildrick
2004b), in part because as Steven Seidman puts it: 'Queers aim less
to normalize gay identities than to free all sexualities from normalizing regulation' (2001: 321–2). His argument can be extended, arguably
without significant distortion, to encompass many of the new directions that are beginning to emerge within disability activism and critical
theory as the question of sexuality moves onto the agenda. The force
of the concern is that it is not simply about a specific or local matter
that can be resolved by broadening the definition of who is acceptable within heteronormative society, but one that requires a challenge
to be made – with consequences for everyone – to the boundaries of
what is sexually permissible. It represents a considerable shift away
from any concept of equality or rights-oriented claims on the sociality, which, as the partial state co-option of gay and lesbian politics
has shown already, tend towards a normalisation that produces social
conformity. Seidman, for example, draws attention elsewhere to the
comforting production of that normalised gay figure as 'fully human, as
the psychological and moral equal of the heterosexual' with the result,
as he puts it, that 'gays should be integrated...as respected citizens'
(2002: 133). The recent moves towards the recognition of civil part-
nerships for gay men and lesbians in both North America and many
European countries has been widely welcomed by pro-gay campaigning
groups, but as Seidman notes, such 'symbolic inclusion into a national
community...has not challenged the meaning of civic inclusion: for
example it has not challenged the...gender conventionality and restrictive intimate and familial norms, that underpin...citizenship' (2001:
323). The question is whether public recognition of the sexuality of
disabled people should be problematised in a similar way.

Clearly there are many significant differences between the gay and
lesbian movement and the disability movement. Insofar as each category represents an outsider status that has both attracted civic exclusion,
and, as I have argued in this chapter, troubled the cultural imaginary,
there are, however, strong reasons to heed the risks that queer theorists
have identified. Though the grasp of normativity may be presented as a
benign embrace, and the desire to bring the extent of sexual expression and desire within the disability community out of the closet is
understandable, there is surely a danger in opening the door to greater
disciplinary control and regulation. The neo-liberal goal of apparently
progressive policy changes – such as those associated with sexuality –
may be to encourage self-regulating citizens who willingly ascribe to

prevalent social norms (Lupton 1999), but that form of governance is but another facet of the governmentality that Foucault (1979b, 1988a) identified as a sign of modern society. And as with the liberal gay man or lesbian who opts for a civil partnership as a form of state-endorsed marriage, the disabled person who seeks open funding of, and support for, her sexual interests is wagering that those pleasures and desires can be accommodated within the social body, precisely because in the end they will cause no disruption. The successful push for equality is conditional, then, on accepting the disciplinary constraints that reposition the hitherto excluded sexuality as simply one difference among many that is encompassed within the normal. What gets left behind – those aspects of sexuality that remain uncontainable within civic society and the cultural imaginary alike – are, as a consequence, all the more devalued and denied.

It is just such issues that have been taken up by Judith Butler (2002) in her own analysis of gay marriage which, in keeping with the general direction of her work, is concerned with the question of which bodies and embodied identities are not just illegitimate, but strictly unintelligible. As she puts it, 'this would be a sexual field that does not have legitimacy as its point of reference' (2002: 232), and she goes on to point out, as did Seidman (2001, 2002), that the co-opting of certain practices to normativity sets up a hierarchy in which those who remain outside are rendered discountable and unrepresentable. As I understand it, to enter into a state of socio-legal recognition is never simply to accrue certain benefits and rights, nor even just to subject oneself to the instruments of governmentality; it is also to project onto those who remain other all the attributes of the sexual self that cannot be legitimated. Butler's question, posed in relation to the arena of gay marriage, is no less pertinent here. She asks: '[D]oes the projection take the form of judging others morally, of enacting a social abjection and, hence, becoming the occasion to institute a new hierarchy of legitimate and illegitimate sexual arrangement?' (2002: 238). Her invocation of Julia Kristeva's (1982) concept of the abject takes us right back to the cultural imaginary which, as should by now be clear, cannot be abstracted from the sociopolitical economy of disability and sexuality. What I have been calling heteronormativity (albeit a heteonormativity that allows for certain sexual variation) is a fantasy that relies precisely on having an outside, a realm of sexual others who are fundamentally inadmissible to the hegemonic and universalising norm.[19] Butler herself is all too aware of the dilemma of needing, on the one hand, to seek relief from the injuries and disenfranchisement of an outsider status and, on the other, of inviting new forms of social hierarchy that foreclose the possibility of alternative sexual arrangements. To gain recognition for some at the expense of others is, as she points out, 'to transform a collective delegitimation into a selective one' (242). It may be that the only extensive commonality among disabled people – who have both variant forms of embodied anomaly and differing sexual desires – is the general devaluation that all may face, but that is precisely the reason why the potential disavowal of some by others is a matter of ethico-political concern.

It is perhaps an artefact of the SMD that any resolution to the social, political, and psychic devaluation experienced by disabled people would be expected to come in the realm of public policy, but that is both to overestimate the power of the legal and social systems to effect change, and to underestimate the significance of the cultural imaginary and its peculiar resistance to anomalous embodiment. Moreover, despite a plethora of undeniable achievements in contesting legal, institutional, and political discrimination, the social constructionist model appears to devalue the phenomenology of embodied difference such that issues of sexual pleasure and desire are rarely incorporated as a primary concern. But my main point here is that the irreducible needs and desires of disabled people cannot be fully addressed, in any case, by a more equitable deployment of the law in its traditional guises, nor by the guarantee of social justice, nor even by sexual citizenship. As I see it, the issue extends to an ethics of the body that is concerned not with principles or regulation alone, but with the phenomenological experience of living-in-the-world through the body. The intersections of self-identity/sexuality/body signal a process of mutual construction, such that a different or altered phenomenology of the body must inevitably figure new forms of sexuality and self-identity that cannot be simply absorbed into or reclaimed to normative standards. The point is that the pursuit of legal, social, and environmental adaptations is not sufficient insofar as it almost inevitably covers over any recognition of the irreducible differences and inherent vulnerabilities – vulnerabilities heightened, no doubt, in the context of sexuality – not just of the disabled body, but of all forms of embodiment. And, as I have been suggesting all along, it is precisely the desire to deny or disavow an internal vulnerability that generates anxiety in the individual psyche and in the cultural imaginary.

Is there, then, any way forward? I offer nothing substantial in terms of policy recommendations – though I hope others will carry these reflections further into that specialist field – but I will continue to assert the value of critique. I am entirely in agreement with both Butler (2002) and Seidman (2001), who have argued variously that public recognition and legitimation are both necessary *and* dangerously circumscribed. The point is neither to seek out the normative fix nor yet to abandon entirely all claims on the sociality; indeed, it is the task of critique to keep that tension in play. Butler suggests that the status of the unthinkable – precisely the status of disability and sexuality – should be valued in its own terms as marking 'a site of pure resistance, a site unco-opted by normativity' (2002: 233), but she recognises that we must go beyond the contentless, and finally futile, celebration of the unrepresented and unspoken. In any case, as I have suggested earlier, we should recognise that silence itself may already be an effect of governmentality and that resistance is not necessarily inseparable from power. What may occur in the arena of disability and sexuality is perhaps somewhat comparable to the discourse that emerged around AIDS where new forms of safe sex practices were both swiftly normalised in mainstream thought and yet remain dynamically transformative of sexual discourse in general.[20] Just as HIV-positive people have been forced to break with heteronormative sexual stereotypes, so too those experiencing some manner of disability are rethinking their modes of sexual becoming. Where full physical sexual autonomy, for example, is not an option for many disabled people, desire can be rethought in terms that do not centre on genital sexuality, or on the goal of self-completion in sexual satisfaction.

That such developments should inevitably trouble the conventions of heteronormative standards, and the cultural imaginary that supports them, is the reason why the conjunction of disability and sexuality, far from being a minority concern, might be a productive force for a queered sexual politics in general. The silencing to which I have alluded throughout this chapter has been broadly construed as having both negative intent and negative effect. Yet even silence may signal something more positive – a potential space for other worlds to emerge – if it were understood as the not-yet thought. There is no pat formula for reconstituting the imaginary but if we do not, at very least, engage with the task of thinking differently, then substantive change is unlikely. The issue, then, is not about gaining sexual citizenship as such, but about the emergence of new forms of embodied selfhood that take account equally of the intersectionality of socio-political context, the meaning of intimacy and the erotic, and the psychic significance of the cultural imaginary. The emergence of sex and sexuality as a matter of concern for disability scholarship and activism alike may have been slow in coming but is now gathering strength. The hope is that it will not allow itself to be caught up in reformist narratives that effectively reproduce existing normativities. I conclude the chapter with the highly pertinent question posed by the academic and disability activist Tom Shakespeare to a conference audience: 'Are we trying to win access for disabled people to the mainstream of sexuality, or are we trying to challenge the way in which sex and sexuality are conceived and expressed and limited in modern societies?' (2000: 163). The contestation of sexual paradigms could go in either direction, but perhaps the most productive way forward is to maintain and negotiate, but never resolve, that tension.

In this chapter I have focused on the socio-political economy of disability, but the task is to explore the problematic of disability, subjectivity, and sexuality through a range of intersecting perspectives that signal not a confusion of intent but the commitment to remain open to new configurations. In the next chapter, then, I shall shift gears and engage with psychoanalysis, another dangerous discourse that many disabled people see as wholly inimical and as signalling a return to the pathologised embodied subject, but which, I will argue, has the capacity to deeply disturb the complacency of those who consider themselves as non-disabled.

4
Sexuality, Subjectivity and Anxiety

The pleasure and danger of sexuality, and more particularly of sexual relations, is a theme that suffuses contemporary society, making clear that the expression of erotic desire must always be accompanied by a certain anxiety. Whilst some of that anxiety is clearly a precautionary response to material risk – the avoidance of unwanted pregnancy, and the contemporary fear of HIV/AIDS, for example, both being concerns that break through specific cultural contexts – the problematic will be addressed in this chapter through an investigation of which psychic factors are at play in the western imaginary. Despite the ubiquity of sexual discourse, the question of who is to count as a sexual subject is contested and uncertain, not just as a matter of practical concern as was apparent in my exploration of social policy concerns, but at the psychological and ontological level too. The western discomfort with many manifestations of erotic desire – that denies or prohibits infant or childhood sexuality, or expresses disgust and attempts to efface older peoples' desire, for instance – is most clearly invoked by forms of differential embodiment that cannot be subsumed unproblematically under the rubric of the normative body. Neither young nor old people are non-normative in their own terms, and yet their difference from the adult body, which is assumed to be the standard for sexual agency and affect – albeit with gender variations – is taken to disqualify them from discourses of pleasure associated with sexuality. Even more disturbing, however, to the point of denial of any sexuality at all, are those modes of embodiment that are both radically anomalous *and* resistant – either projectively or retrospectively – to normative recuperation. The category of congenital or early onset disability[1] is surely paradigmatic in that its exclusion from the very notion of sexual subjectivity is so sedimented that it is scarcely challenged on an empirical level and even more rarely interrogated as to the nature of its psychic underpinnings. If the contestation of social policy has been, and remains, a struggle, how much more so is the unpicking of that which is, by definition, subconscious? In this chapter my purpose is no longer to enquire into the multiple ways that disabled people are denied sexual subjectivity, but to address the question of what is at stake in the cultural imaginary that requires such a closing down of possibilities. Following on from earlier chapters which have already explored aspects of the cultural, historical, political and legal devaluation of corporeal difference, I want here to contextualise the putative threat of anomalous embodiment in terms of the specific intercorporeality of the sexual relation.

Before moving on, however, it is necessary once more to remind ourselves that disability takes many forms not all of which are written on the body. At a broad level it is all too easy to conflate what are variously called developmental or learning disabilities with physical disabilities, and although a similar analysis applied across categories may be productive, it may also obscure highly cogent differences. And whilst I am somewhat wary of categorical distinctions between the mental and physical, it is true to say that my own work – like that of the majority of other disability theorists – is more firmly focused on corporeality as a mode of embodiment.[2] With these provisos in mind, I am confident that while some of my observations are applicable more generally, the focus in this chapter is firmly on what is *visibly* anomalous to a degree that challenges normative expectations of the human body. To Lennard Davis, physical disability is 'a disruption in the sensory field of the observer' (2002: 50), and while I agree wholeheartedly with that assessment, I am more concerned here with the consequent disruption to an observer's psychic field. It should not necessarily be assumed, however, that normative expectations are limited to observers who are themselves classed as non-disabled. While the greatest problem for people classed as having some disability may spring from the disavowals of that majority, the *self-understanding* of every one of us – however we are positioned – is inevitably deeply imbricated with the cultural imaginary to the extent that, like people of colour, gays or women, disabled people may both accept and resist the dominant discourse. In bringing the insights of psychoanalysis to bear on the nature of the cultural imaginary itself, I am clear, at the same time, that much future work and discussion is needed in order to better understand the difference that variant morphology makes at a psychic level to the development of self-image, and how that feeds into socially situated issues of self-esteem.

What is at stake, then, is not some simple opposition between cate-gorical groups with and without socio-cultural capital, but a discursive context in which a range of people – those who are disabled, those who are normatively embodied, disability scholars, and activists and allies alike – are all caught up in the interplay of barely recognised forces. Nonetheless, as I note later, it should never be forgotten that the sig-nificance of the interventionary potential and limits of psychoanalytic discourse plays out differently across those varied groupings. There are powerful reasons why psychoanalysis plays so little part in disability studies other than as an exemplary referent of oppressive forces, but the fear that it must depoliticise by focusing on the mental state of the individual fails to acknowledge its other dimensions. My own strategy seeks to uncover the psychic work that supports the cultural imagi-nary, both as it operates through individuals and as it symbolises the normative structuring of society. In any case, unlike the trajectory of most conventional accounts, what I am attempting is to set out not an over-attenuated explanation of the state of mind of disabled people, but rather to gain some insight into the apprehensions of those who count themselves part of the 'ablebodied' majority. For the purposes of this chapter, then, my intention is to trouble and disrupt dominant power relations, not from the position of those who are othered, but, as it were, from inside normativity itself. To direct my reflections primarily – though never exclusively – towards people who are non-disabled, how-ever, does not mark a belief that the members of any singular category are uniquely implicated. To put it starkly that those in the mainstream are the problem should not be taken to entirely limit the field of enquiry, given that we all share certain psychic origins. Moreover, to locate the problematic within a nexus of socio-cultural forces, rather than as a matter of personal construction, in no order undermines its psychic inflection, for what is at stake is the critical intersection of material expe-rience, representation, and psychosomatic symbolisation into which we are all differentially drawn.

In the West, all forms of differential embodiment are highly pro-ductive of normative anxiety in as much as they threaten to overflow the boundaries of what Kristeva calls 'the self's clean and proper body' (1982: 71). Despite the post-Cartesian entrenchment of the notion that the self-possessive inviolability of the bounded body grounds the autonomous subject, most of us are both consciously and subcon-sciously engaged in ongoing strategies that provide protection against the putative dangers of encroachment, even engulfment, that other bodies seem to pose. One consequence of seeking to maintain the

illusion of the separation and distinction necessary to the sovereign subject is that all encounters between self and other are potentially risky,
and must be negotiated within a strict set of normative rules and regulations that construct the parameters of safety and danger.[3] In such a
context, the sexual relation itself and the operation of desire, as that
which extends beyond the self to the other, is always a locus of anxiety.
Although, as Liz Grosz puts it, 'erotic attachment induces a *voluntary*
reciprocity in vulnerability' (2006: 193, my emphasis), it still remains a
potent point of disturbance to the normativities of everyday life. The
risk of sex is not, however, limited to the potential loss of self in the
other, but carries too a threat to corporeal integrity, specifically, as I discuss later, of regression to the fragmentation associated with infancy. As
Alphonso Lingis remarks of the orgasmic body:

> Is it not a breaking down into a mass of exposed organs, secretions,
> striated muscles, systems turning into pulp and susceptibility? The
> orgasmic body is… the body drifting toward a state on the far side of
> organization and sense…
>
> (1985: 55–6)

In consequence, sexuality both invites the apparent protection of
heightened disciplinary techniques, and continually breaches them.
And where sexual discourse characteristically links pleasure and danger
in the erotic, it is not simply as opposing or alternative courses
or outcomes. Pleasure and danger are indivisible and figure an irreducibly intertwined dimension in the very structure of the sexual
relation. Nonetheless, our investments in a cultural imaginary of self-
determination work to occlude the uncertainty of all sexuality by
imposing an impressive system of legal, cultural, and moral constraints
that construct categories of the licit and illicit that are supposed to operate to contain or eliminate perceived risk. But what is it that determines
which behaviours, or indeed which bodies, are classed as illicit?

My argument is that if any coming together of bodies, and more
specifically the intercorporeality of much sexuality, is encompassed
within an implicit anxiety about the loss of self-definition, then that
anxiety – which operates within us all – is at its most acute where the
body of the other already breaches normative standards of embodiment.
Insofar as the other in its alterity is always a possible threat to the
integrity of the self, then the radically different other – the one who
is not like me, who fails to observe the same boundaries – is doubly so.
The illicit, then, concerns a potentially contaminatory threat to the self,

an interchange not between self and other where each is in a familiar category, but between strangers where the substance of what passes across boundaries – be it sensation, emotion, or material – is unknown. For disabled people, who are routinely positioned as outsiders, to be faced with restrictions on their own sexual pleasure with the implication that it is somehow illicit, or the denial even that their sexuality could be significant at all, is a habitual affront. It speaks not to any engagement with the realities of lives that in their differential embodiment have differential possibilities of expression and connection, but at least in part to the fear occasioned when bodily self-control seems to be, or is, compromised. Despite the widespread misconceptions about disabled people's sexuality which permeate not only lay opinion, but social and welfare policy, recent empirical evidence supports the view that despite sometimes necessary variations in active expression, there is little distinction to be made in the levels of sexual desire (Nosek *et al* 2001; Shuttleworth 2002; Vansteenwegen *et al* 2003). At the same time disability activists and scholars – and particularly those working within queer and feminist theory – are increasingly problematising the conventional parameters of sexuality in order to explore non-normative constructions of sexual identities, pleasures and agency that more adequately encompass multifarious forms of embodied difference.[4] For the reasons already discussed, it is not that people identifying as disabled are exempt from the cultural imaginary that mobilises anxiety, but that sometimes their practices must, perforce, break through the boundaries of everyday convention. Far from being a disadvantage, it could be read as both innovative and productive. As Tom Shakespeare comments, 'non-disabled men have things to learn from disabled men, and could profitably share insights into gender relations, sexuality and particularly issues of physicality and the body' (1999: 63).

The manifest insecurities which I am addressing, then, are not primarily those of the disabled person. They belong instead to the one who though considering herself indisputably within normative parameters, nonetheless experiences an inherent anxiety attributable to the uncontrollable fluidity – a certain messiness – of the sexual relation, and who is further undermined by contact with the unpredictability of a body that does not behave as her own. Though the ability to draw boundaries between self and other may be temporarily suspended in *any* sexual encounter, those distinctions are normally recuperable. Given the further element of an encounter with the strange(r) other, however, that ability is put in permanent doubt. It is as though the subject's own self-control were at stake. One major consequence is that the sexuality of disabled people may be seen subliminally as a threat, both for the normative majority, and – because none of us is outside the dominant discourse – in the disability community itself in the form of what amounts to self-censorship. The resulting unease is evident equally when such sexuality is effectively silenced, and when it is spoken. In a very common complaint that public representations veer between the asexual and the hypersexual, Alison Kafer notes:

> there has been an excited discourse around disabled people's sexuality as inherently kinky, bizarre and exotic. Medical and popular assumptions that people with disabilities are asexual contribute to the discourse about sexuality and disability – while the sexuality of disabled people may be denied in these conversations, it is being denied loudly and repeatedly, not silently.
>
> (2003: 85)

What Kafer explicates here parallels precisely Foucault's account of the discursive coming into being of homosexuality through the very attempt to cover over the practices thus named. Nonetheless, the category that is simultaneously produced and denied cannot simply emerge as positive by ensuring – as some too hasty accounts seem to assume – that more sexually affirmative representations are put in place. Aside from the Foucauldian exposition of the eternal play between power and resistance, which I shall not consider further here, another, less-considered element works against any easy resolution.

The devaluing of sexuality in the context of disability is so powerfully entrenched in western culture and social policy, so bound about with unspoken anxieties at any sign of the eruption of such sexuality, that it suggests that the very lack of adequate recognition should be characterised as a *disavowal* that requires not simply a socio-cultural analysis but an enquiry into its psychic origins. It is not that socio-cultural organisation and behaviour can be read directly off a psychoanalytic analysis, but that the apprehensions and impasses that surround everyday practices are grounded in a series of unconscious motivations stemming from early infancy and persisting throughout adulthood. It would be difficult to overstate the complexity of psychic development, and as yet, the existing literature on disability and sexuality using that perspective is disappointingly sparse. The reason for its relative absence is not, however, a reluctance to engage with intricate theory, but has several other explanations. At very least the gap is explained in part by the emphasis given by the dominant social model of disability to more material socio-political concerns that are susceptible to rights discourse. The model, as I have already outlined, has little or nothing to say on the subject of sexuality, and even less has a place for the question of desire. A further inhibiting factor is surely that even when sexuality is deemed to be of high significance, it is often read through a broadly Foucauldian model that gives little credence to psychic factors. As Foucault himself puts it, what matters is the relationship between what we actually do, are obliged, allowed, and forbidden to do in the field of sexuality, 'and what we are allowed, forbidden, or obliged to say about our sexual behaviour.... It's not a problem of fantasy; it's a problem of verbalization' (1997: 125–6). In short, it is not a perspective that encourages enquiry into the full complexity of our motivations. Finally, and I shall return to this later, the discourse of psychoanalysis that gives the clearest account of psychic matters is just one of those domains that, because of its imbrication with issues of mental capability, has represented over the last few decades a no-go area for disability activists and theorists. The dangers of evoking the putative link between physical disability and mental incapacity are clear, but it is precisely to disarm those threats that an engagement with psychoanalysis on its own terms becomes so important.

My primary concern is with bodies, and while it is undeniable that disavowal is as forcefully operative in the context of developmental as in physical disability (and perhaps in some scenarios even more so), for now my analysis is largely limited to visibly anomalous embodiment, and I hope others will tease out the limits of its wider applicability to non-visible disabilities. Nonetheless, I should like to take a moment to reflect further on how visibility itself plays into instantiating the (illusory) divisions – between physical and intellectual anomaly as much as between ablebodied and disabled – that socio-cultural norms rely on. Hannabach (2007) suggests that it is *primarily* the visibly anomalous that both defines bodily difference and simultaneously troubles the imaginary, and certainly that is one interpretation that this chapter would seem to support. Nonetheless, there are, I think, two questions to consider. First, we need to ask whether what is visible does in fact disturb the imaginary more forcefully than other forms of difference. Might it not be the case that it is just easier to talk about the visual precisely because it is *less* disturbing, insofar as there is always a gap, an interval, between perceiver and perceived? That is not to deny the reversibility that Merleau-Ponty (1968) speaks of, but to suggest that a certain distance can be maintained, at least by those in a normative state of mental well-being. When it comes to affective anomaly, or more fully what we would understand as developmental difference, the putative threat to our psychic stability is perhaps even greater. As someone who has been engaged with critical disability theory for many years, I am continually struck by the reluctance, both of myself and others, to move beyond the question of visible anomalies. My own failure, as yet, to adequately address the issue of sexuality and developmental disability, for example, is not, I suspect, a simple matter of assessing where the greater relevance to my project of deconstructing corporeal and ontological anxiety lies, but more in the nature of a resistance to the disorders of mind. It is notable that on the infrequent occasions when the sexuality of people with developmental differences is discussed, it is very much in the mode of an *overview* – the 'safe' metaphor is telling – of those 'others'. My second and related point is that I find myself wondering about the status of the visible in psychoanalysis itself where it seems to be both privileged, as in Lacan's mirror stage (albeit in the mode of misrecognition), and subsumed by language, once the entry into the symbolic is effected. What this should remind us of is that language is always saturated with visual metaphor and perception itself is discursive. This suggests the fragility of categories of visible difference, but it is unclear to what extent psychoanalysis contests the nature of those differences. It would be strange if psychoanalytic theory could offer no insights into that form of normative anxiety.

My argument is that in relation to the potency of the cultural imaginary, and to what is effectively the question of the other, psychic dimensions can provide a number of convincing accounts that throw up many potentially fruitful avenues of enquiry.[5] In this I follow Jacqueline Rose who has argued convincingly in another context that the socio-cultural realm should not be addressed 'as if it were free of psychic and sexual processes, as if it operated outside the range of their effects' (1993: 41). I shall concentrate on psychoanalysis and on the effect of *unconscious* prohibitions on desire, rather than simply examining the way in which thwarted desire plays out in the realm of the psychosocial around particular forms of embodiment. And the invocation of the unconscious here does not intend, it should be stressed, the repressed and unspoken locus of the individual psyche, but rather that which structures the possible manifestations of embodied desire itself. Although the discipline of psychoanalysis is often criticised for invoking a certain ahistorical grounding for its central concepts, that should not disqualify its insights. There is nothing in either the Freudian or Lacanian canon that would deny the specific ideological investments that shape the particular manifestation of psychic concerns, and while the infiltration of such investments into psychoanalytic discourse itself is never discussed as such, a reflexive approach will always bear that in mind. Freud clearly understood the imbrication of the social and the psychic, in which the body, and especially sexuality, acts as a privileged mediator; while as Malone and Kelly note: 'Lacan's polemics about ego psychology, social engineering, the mediocrity of psychology all point to his sensitivity to the permeability of the psychoanalytic closet to socially regnant ideologies' (2004: 26). Though it is not the primary purpose of psychoanalysis to uncover, for example, the variously raced or classed specificities of its subjects, it can provide the tools whereby sense might nonetheless be made of the differential instantiation of what it takes to be individually transhistorical and transcultural psychic determinants. The Lacanian turn to the constitution of subjectivity clearly opens up the field to a reading that takes account of the mutually determining effects of gender, culture, and race, for instance, and – though it is rarely remarked – (dis)ability.

Let me begin, then, by developing the psycho-cultural significance of the Lacanian infant body that displays in the early months after birth an inherent dis-integration and frustrated mobility, and yet emerges subsequently as a coherent sexed and sexual subject in the Symbolic. Leaving aside the effects of physiological development, what, we can ask, has been repressed, or cast aside, in order to achieve that putative unity and order? More importantly, which forms of embodied subjectivity cannot be countenanced, and how might that relate to the disavowed sexuality of disabled people? In opening up this particular line of enquiry, I do not expect it to resolve the questions I have posed (which are perhaps not even the right questions), but rather to suggest ways in which the 'why' of the problematic rather than the 'how' of its effects could be investigated. Regardless of how any one of us is embodied, the refusal to countenance what lies beyond the normative, and the consignment of the sexuality of some people to a dangerous discourse, is an effect of the parameters that put limits on the sexuality of us all. The task then is both to retrace the constitution of the normative sexual subject and to reclaim other modes of sexual becoming. The psychoanalytic narrative proposed by Freud, and reconfigured and reiterated by Lacan, is deeply disruptive of the supposed security of the subject in the broadest sense, and more particularly so in that the inherent excessiveness of sexuality risks always splintering the imaginary unity of the bounded ego. Psychoanalysis lays bare, moreover, the intimate connection between desire and anxiety, which for Lacan at least are both necessary constituents of the subject. To characterise desire in terms of lack as Lacan does is, however, by no means self-evident and I shall in due course suggest that his schema should be read against the insistence of Deleuze and Guattari that desire is always productive and mobile. It is not that the effects of Lacanian desire are unproductive or static, for lack itself mobilises a narrative logic in which the subject compulsively seeks completion in sexual practices. The point is more that such desire is inward-focused rather than expansive. Nonetheless, in a very positive take on the potential of Lacan's account, Tim Dean makes the strong claim that in Lacan, '(the) excess of meaning called the unconscious generates desire as a multiplicity of possible connections' (2000: 250). Perhaps the apparent Deleuzian break with psychoanalysis is not so radical after all. I shall propose an alternative Deleuzian take on disability and sexuality in the Chapter 6, but for now, let me review the Lacanian understanding of how the early infant emerges as a subject.

According to the most familiar formulation outlined in the essay of the same name (Lacan 1977b), the infant – prior to the 'mirror stage' – experiences himself or herself only as a body in bits and pieces, as fragmentary and discontinuous rather than as whole and integrated. It is the experience of what Lacan calls '*le corps morcelé*', a body that is no more or less real than the conceptually unified corpus that will eventually take its place. What is undeniable is that the infant, far from achieving the self-determining normative standards that characterise adult life, is immersed in 'motor incapacity and nursling dependency' (1977b), a state in which the mother is an all-powerful figure, but one not separable from the infant itself. As is well known, Lacan figures the rupture in the infant-maternal dyad as the moment at which the growing child comes to realise a mirrored image of his own body as putatively whole and separate, and thus is enabled to disavow or cover over his actual incapacity and disarticulation in the assumption of what Lacan calls 'the armour of an alienating identity' (4). Yet that emergence of an integrated self is based from the first on a misrecognition, an identification with an exteriority, that serves to inaugurate both *an imaginary anatomy and a phantasmatic self*. I stress this element as it will be crucial to the extension of Lacan's account to my own reading of disability. Equally critical is the claim that it is not simply physical disunity that is cast aside, but a startling series of negative images to which, Lacan asserts, the infant is exposed in the imaginary prior to the illusion of wholeness: 'images of castration, mutilation, dismemberment, dislocation, evisceration, devouring, bursting open of the body, in short...*imagos of the fragmented body*' (1977a: 11). These are surely highly significant, for they will remind us not so much perhaps of the disabled bodies that we actually encounter or experience for ourselves day to day, but certainly of the socio-cultural fantasies that have always surrounded disability, particularly in historical representations of the monstrous body (Shildrick 2002). These are the forms of embodiment that must be suppressed in order for the child to achieve the stability and distinction that mark out the normatively embodied subject, although as Lacan makes clear they re-emerge in the subject in the form of aggressivity.

What Lacan's scenario seems to imply is that any body that should persist in its manifestation of dis-integration and disunity – paradigmatically any visibly disabled body – may become the repository of both corporeal and ontological anxiety, and even violent response. The phantasmatic sexed and gendered subject who emerges in the Symbolic – where desire for the mother has been superseded by the Law of the Father – is always fissured by its own inaugural *méconnaissance*, and deeply insecure in its apparent normativity. Its hold on order, control and self-determination is fragile and uncertain, maintained only by strategies that hold at bay those others whose own corporeality re-awakens intimations of a fundamental dis-organisation, and lack of self-completion. It is here that the western wariness about, or even aversion to, touch is surely rooted, for above all the normative subject must avoid the indeterminate physicality of intercorporeality. The scopic drive, heralded by the mirror stage and remaining dominant throughout life, is the primary means by which the subject distinguishes itself from its others. But if the specular interval between self and other is what serves to protect the illusion of singularity and corporeal unity, then any point of contact, any tactile encounter, threatens a loss of differentiation. Where in addition the body of the other cannot be subsumed under the rubric of identity to the selfsame, then its putative danger to psychic security is greatly accentuated. It may appear overwhelming – in both a metaphorical and literal sense. The enduring difficulty is that although the otherness of such threatening corporeal forms must be repudiated, that otherness is not simply external to the subject, but always recalls the self's incompletely repressed experience of infantile dis-organisation. The relation of the anomalously embodied other to the subject takes on all the irresolvable ambiguity and disruptive potential of the abject: neither wholly part of the subject, nor safely outside; disavowed, yet disturbingly familiar (Kristeva 1982, Shildrick 2005b). Indeed, Kristeva (1990) herself makes the links with the Freudian concept of the uncanny as that which irrupts unexplained into consciousness when it should not appear. Lacan does not speak of the abject as such, but given that he identifies Freud's 1919 essay *The Uncanny* (Freud 1962) as a grounding text from which to theorise anxiety (Harari 2001), the connection is assured.

In contradistinction to other psychoanalytic approaches to the phenomenon, Lacan insists in his unpublished seminar on anxiety not only that anxiety is 'not without an object', but quite the reverse: that its appearance signals the too-closeness of an other who is excessive to representation and who threatens the symbolic order (Harari 2001). And Richard Boothby specifies the same point: 'Inasmuch as it is formed on the basis of a unifying perceptual *Gesalt*, the ego is liable to anxiety in fantasies of the fragmented body, or *corps morcelé*' (1991: 143). The threat, in other words, is firmly directed towards the psychic structure of the self. The issue then is not that the object of anxiety fails to appear, or has been lost, as Freud sometimes understands it, but that on the contrary what should be lacking is perversely present. Indeed it closely figures Freud's own characterisation of the uncanny. In the essay of the same name, Freud refers to phantasies of '(s)evered limbs, a severed head, a hand detached from the arm... feet that dance by themselves' as *unheimlich*, particularly in as far as 'they prove capable of independent activity' (2003: 150) These too are *imagos of the fragmented body* which not only reappear when they should not, but threaten to usurp subjective agency. In Lacanian terms, such presence must be of crucial significance for the staging of the symbolic subject, given that lack is a necessary constitutive element of that subject. The danger, then, lies precisely in lacking lack. So, for Lacan, what is it that appears? This is how Roberto Harari glosses Lacan: 'The object that provokes anxiety... (is) the desire of the Other, as the Other requires that the subject erase its borders, handing itself over to it in an unconditional manner' (2001: 75); and again: 'Structurally speaking, anxiety... suspends the imaginary mapping that intuitively recognises the difference between an inside and an outside' (2001: 162). In other words, faced with the reappearance of its pre-subjectival phantasies that should have been banished – faced in substantive terms with the body of disability – the subject is endangered by the putative failure of its own boundaries of distinction and separation. And Lacan makes it quite clear that the threat of such disarray and fragmentedness is not some contingent aberration from the normal experience of the human being but is intrinsic to every one of us (1988: 169). Accordingly, the trigger that reopens a conscious experience of anxiety, or more seriously of outwardly projected hostility, need be nothing more extraordinary than a corporeal encounter with disability.

Given the potential of such an approach to offer some explanatory account of the relation in the imaginary – which as Lacan notes is always structured in binary terms – between disabled and non-disabled, why is it that most disability theorists remain reluctant to investigate the insights of psychoanalysis? The explanation is surely that the rejection of the medical model, which dominated the understanding of disability until relatively late in the twentieth century, is so strongly entrenched in contemporary disability politics that any perspective that seems to entail a return to individual pathology is to be avoided. In common with many other theorists of 'minority' interests such as those associated with gender, race, and sexuality, disability scholars have been rightly suspicious of a psychoanalysis that has a history of complicity with, and active construction of, models that pathologise all forms of embodiment that exceed the norms of morphology, identity, and desire. Moreover, psychoanalysis, whilst not exactly 'blaming the victim', is widely understood to individualise the experience of oppression rather than attending to structural, and therefore political, determinants. What that accusation sets up, however, is a very limited either/or understanding of the issue of oppression that entirely misses the dimension with which I am concerned. As Hannabach puts it: 'Psychoanalysis tells a story, but it is not the story of an individual person – instead, (it) is a story of a cultural imaginary that is controlled by no one but that structures every subject' (2007: 257). In such a way, psychoanalysis offers disability theorists a means of interrogating and contesting the unacknowledged foundations of the normative narratives that both limit some and enable others, although never without the potential of categorical confusion. As an enquiry into the sexual imaginary, it is concerned above all with uncovering the structural prohibitions that disavow some marked bodies as potential objects – or subjects – of desire so as to establish and sustain the non-disabled body both as desirable and as a proper desiring subject. And because psychoanalysis makes clear that the normative subject herself is constituted precisely in the mode of fragmentation and loss, the possibilities of opening up what might be called 'the other side of underneath' might seem self-evidently attractive to disability theorists. Nonetheless, the confusion between an approach that psychologises – and in a strong sense privatises – the experience of disability, and a psychoanalytic approach that enquires into the cultural imaginary is widespread, and has been successfully challenged very rarely.

There are, however, some noteworthy exceptions.[6] Robert Wilton's sophisticated essay that analyses the way in which disability figures a symbolic substitute for the trope of castration in Freud and Lacan offers an important addition to the field of critical disability studies, and yet it is almost unique in its development (Wilton 2003). Significantly, Wilton makes consistent and strong links between the social expression of disability as a 'tragic loss', the lack that it stands in for in psychoanalytic discourse, and the generation of a psychic anxiety that precludes the positive expression of disabled sexuality. In more widely known disability scholarship, Lennard Davis' brief excursus into Lacanian theory is equally radical when he points out that in the specular moment of encountering the disabled body, 'the *moi* is threatened with a breaking up, literally, of its structure, with a reminder of its incompleteness', and that the 'normal' body 'is in effect a Gesalt – and therefore in the realm of what Lacan calls the "imaginary"' (1997: 61). Davis' account remains relatively underdeveloped, nonetheless, and he makes no move into the area where disability intersects with sexuality. In contradistinction, it is my own strong contention that once the problematic is moved away from the abstraction of theory and posed within the context of everyday possibilities, it becomes clear that it is the encounter with an anomalously embodied other positioned within the arena of sexuality that is the most threatening and disruptive. The opening up of the subject to a certain degree of the uncertainty and risk that is always inherent in the intercorporeality of the sexual relation is significantly intensified by the operation of corporeal difference. The psychic underpinnings of the disavowal of the very possibility that the disabled person should be sexually active, sexually identified, and sexually engaging become all too clear.

In order to more fully appreciate how the psychic register plays out in the material world of differential embodiment, it is important to note that anxiety is never merely negative but is generated as much by fascination as by threat. The fraught relations of pleasure and danger are always complicated by the ambiguous nature of desire itself. The point is that as a psychic response in the unconscious, desire must always represent – in psychoanalytic terms at least – a failure of satisfaction, a lack of self-completion that exists only because the object of desire has already been lost. If, then, desire is structurally articulated with lack, it will always be mired in the tension not only of opposing sensations (erotic excitation and its falling short), but of identity formation itself: the subject, we should recall, cannot come into being without the element of lack. The goal of desire – what Lacan (1981) calls the *objet petit a* – is caught up in an endless series of displacements in the place of the other that substitute for the originary loss of the first love-object, the mother. In terms of its psychic meaning, that loss is the necessary consequence of the symbolic castration effected by the Law of the Father that cuts the child off from the dangerous plenitude and *jouissance* of the maternal–infant dyad and instead promotes the advent of the speaking subject.[7] In some accounts that do no justice to either Freud or Lacan (Minsky 1996), the seemingly nostalgic fantasy of a 'return to the womb' is used only to figure precisely what is desirable, but that is to overlook the anxiety-provoking threat that such a reintegration implies. Because desire, in the sense of a demand that is beyond satisfaction, is operative only in relation to the child's accession to independent selfhood in the Symbolic, it is paradoxically both a desire for an *imaginary* self-completion in the desire of the other, and a longing for the indistinction of intercorporeal dependency. With regard to the latter, desire, in effect, is founded on and directed towards the very thing that must be repudiated in order to achieve a putatively stable identity: the subject, then, must either turn away or risk dissolution. As Robyn Ferrell succinctly notes: 'Repression can be seen as a process of distinction' (1996: 16). But does what is repressed here – desire for/of the mother – finds its counterpart in other repressions that equally speak to an intercorporeality that is both seductive and threatening?

Now clearly desire in the sense outlined is by no means coincident with what is meant by adult sexuality, though that is precisely where the striving for, and failure of, satisfaction most usually plays out. What is significant from the point of view of understanding the socio-cultural response to the disabled body in a sexual context is the way in which the practices of normative sexuality are assumed to develop. If, as psychoanalysis proposes, they originate in the infant's initial experiences of erotic pleasure in its own body – pleasure that is set in motion by close maternal care in the mode of breast-feeding, cleansing and rubbing dry the genital organs, by body manipulation in all its forms, and so on – then adult sexuality itself, as the major trajectory of desire, is permeated with a nostalgia for the fragmented, incomplete body, the body, in other words, that is intrinsically dependent on another.[8] It is, at the limit, a lingering desire for a phantasmatic identification – the *corps-à-corps* – with the maternal figure. What can also be deduced in the light of such pre-subjectival underpinnings is the further and somewhat contrary insight that all desire is deeply rooted in narcissism; pleasure is centred on the (unformed) self in a mode that must stand against the operation of any *intersubjective* desire for the other *as* other. It is clear that the psychic interconnections that bring together desire and identification are highly complex, and, indeed, that they are always potentially destabilising to the embodied subject. What is at stake here is not only that the putative mother is the focus and grounding of both desire *and* disavowal, but that whilst any intimation of a return to the dis-organised *corps morcelé* is a matter of psychic danger, its half-remembered pleasures also signify a seductive promise. On the basis of such paradoxes, which continue to multiply and thread throughout the domain of sexuality, could we not speculate that the rejected body of disability is at once what figures the repression of the self's own fragmented body, and what figures a state of sexual satisfaction? Could it be that the normatively bounded and embodied self experiences a psychic pull precisely towards that which it overtly rejects.

What I am suggesting is that for the subject in the symbolic order the disabled figure may represent, in a way analogous to the mother herself, both a forbidden and desired other,[9] the locus of a lost sexuality that lacked nothing. The characteristic response to people with evident disabilities exhibits just such a split in that alongside a normative imperative to devalue or silence their sexuality, there is also a highly evident strand of voyeurism – and in some cases identification – that spills over into a fetishistic focus on disabled bodies precisely as sexual.[10] Moreover, the differential responses are by no means mutually exclusive. On the one hand, then, there may be not simply a refusal of recognition, but something approaching disgust, a phenomenon that cannot be simply opposed to pleasure, for in it, as Menninghaus (2003) notes, the ambivalence of attraction and repulsion is always already at work. Indeed, disgust 'disguise(s) a repressed or rejected...pleasure, rather than the total absence of any relationship with pleasure' (2003: 36). On the other hand, a very slight shift of emphasis accomplishes the move from disgust to embrace, and the endorsement of what might otherwise be disavowed. A good example of the latter – though the operation of pleasure is only one of the many dynamics in play – would be the growing number of internet sites intended for self-named amputee devotees.[11] In the disability movement, devotees are the subject of contentious debate largely around the issue of power relations, but what is interesting from a psychoanalytic perspective is the extent to which even a putatively obsessive fixation calls into question the rigidity of normative sexual development.[12] More generally, the doubled circulation of both fear and fascination that is so often evident with regard to anomalous bodies (Shildrick 2002), and that more specifically here marks the context of disability and sexuality, speaks to an inherent instability and vulnerability in a normative order that relies, above all, on repression. And as Freud himself recognises, repression is both necessary for the emergence of 'civilised' society, and the source of its insecurity (1962).

As a preliminary and limited enquiry into the interior sense of pleasure and danger that frames the cultural imaginary, the model so far proposed has a persuasive explanatory power, but analogous to the Lacanian account of the feminine which has long troubled feminism, it seems somewhat gloomy, in this case for those people who are disabled. There is no doubt that all of us have a responsibility to recognise that the constitution of the normative itself warrants thorough investigation, and that a Lacanian-based exposition of normative anxiety undeniably goes some considerable way towards an explanation of troubled disabled/non-disabled relationality. The problem is not that the model disturbs a complacent majority, but that it scarcely yields a positive account for those who wish to celebrate sexuality and sexual expression *within* the context of disability. In other words, although the explication of the sexual dynamic seems compelling, disabled people themselves may nonetheless reasonably question the *utility* of the psychoanalytic approach.[13] Indeed, if instead of seeking to understand the socio-cultural unease that underlies the silencing of such sexuality, the focus were to shift away from the normative to the anomalous, it is as yet unclear whether the psychoanalytic model could escape the burden of the apparent impasse into which non-normative sexuality is driven. It is not that the ambivalence that subtends all expressions of desire can be put aside, but that the model may fail to provide the ground on which the hierarchical structure of the disabled/non-disabled binary – and the sexual expression 'proper' to each category – could be contested. The urgent question is whether there is space within the model for things to be otherwise. Given the apparently inherent phallic masculinity and heteronormativity of the model, it is an issue of how desire can be figured as belonging to an excluded group in which just those attributes may, for functional as well as preferential reasons, be at stake, and whether the sexuality of those who are disabled could be adequately encompassed within the analytical framework laid out. The danger is that any attempt to configure the sexual desire of disabled people on the narrow ground of (hetero)normativity risks fixing it as inherently defective and incomplete. There are troubling conceptual questions too: does, for example, the mirror stage, as the threshold of the self–other relation that grounds mature sexuality, take on the same significance for the infant whose actual bodily dis-organisation exceeds and outlasts any illusory reflection? If the *ethical* task of enquiry is to instantiate a more positive mode of thinking the sexuality of disability itself, then the

Lacanian model may be simply ineffective, and the exploration of a less apparently deterministic alternative – a Deleuzian model for example, which I shall explore in a later chapter – might prove more productive.

To return to my question, however, it is necessary to ask whether the Lacanian, or Freudian model from which the former derives, could encompass the possibility of being otherwise, without that signalling a failure of psychic development. Let us for a moment revisit the crucial stage at which the infant acquires its sense of itself as whole and separate, the point at which the possibility of sexual subjectivity – indeed subjectivity as such – is opened up. As Lacan recounts it, the moment is one of a 'triumphant jubilation and playful discovery that characterize...the child's encounter with his image in the mirror' (1977a: 18), an encounter that appears to set the scene for a disavowal of the early infant body as it now figures *retrospectively* in the imaginary as wounded and dis-organised. It is not the newborn in its first months who feels disturbed in response to a sense of its own disunity, for its incapacity is covered over initially by the irreducibility of the maternal–infant dyad; rather, at and after the mirror stage, it is the emergent *subject* who experiences 'the succession of phantasies that extends from a fragmented body-image to a form of its totality' (Lacan 1977b: 4).[14] As Debra Bergoffen points out, the infant

> reads its original experience of embodiment through the lens of the *later imago* to call it fragmented and unco-ordinated rather than exuberant and polymorphous. The adult, orienting itself round the legacy of the imago, reads its experience of the ambiguous body through the lens of the ego to call it threatening rather than exhilarating.
>
> (2000: 103, my emphasis)

Now although what Bergoffen is addressing explicitly at this point is the question of how the inauguration of gender is linked to the desires at play in what she calls 'the metonymic-imaginary game' of the mirror, her remarks invoke other dimensions. As I have already outlined, it is surely possible to surmise that what is at stake here is the formation of a cultural imaginary in which the disabled body – precisely in its incompletion, ambiguity and lack of co-ordination – is rejected or disavowed. But is there any chink in the armour of the normative subject – 'the armour of an alienating identity' in Lacan's words – that could make it otherwise? Is the anxious, yet – insofar as fascination with that other remains – psychically incomplete, putting aside of the body that lacks wholeness and unity an unavoidable step for the emergence of a sexed and gendered subject, or simply one possible trajectory among others? The wider issue is whether the normative framing of the psychoanalytic account represents the limits of its intelligibility, or simply a contingent closure that can be opened up to a different reading. Might a return to Freud provide some answers?

Unlike Lacan, for whom the early phase of infancy indeed appears beset by negativity and danger, Freud presents the same lack of coherent impulses in a more positive light. In *Three Essays on the Theory of Sexuality* (1962) he sets out his theory of polymorphous perversity in terms of it being the originary capacity of the infant to take sexual pleasure in every part of his or her own body. Before the staging of what he sees as the ideally sequential development of erotogenic zones of the body – the oral, the anal, and the genital – Freud's proposition is that the body's surface is undifferentiated in its facility for sexual excitation. In origin, then, sexuality is highly versatile and plastic, and subsequent libidinal investments in *specific* loci are the result not of some natural instinct, but of an intensification of affect and meaning, an intensification that is both contingent and regulated. The turning away from all but sexually differentiated genital sexuality – in the face of 'shame, disgust and morality' – is a development that fixes adult sexuality as the outcome of a process of repression about which Freud himself is increasingly ambivalent. Even at the time of the relatively early appearance of the collected *Three Essays* in 1905, when Freud is cautious in his approach to so controversial an issue, he draws the conclusion from his study of childhood sexuality (albeit one based on the analysis of adult patients) that:

> a disposition to perversions is an original and universal disposition of the human sexual instinct and that normal sexual behaviour is developed out of it as a result of organic changes and psychical inhibitions occurring in the course of maturation.
>
> (1962: 97)

In other words, the distinction between perverse and 'normal' sexuality is far from clear-cut, and moreover, the latter cannot be thought without the former. What Freud meant by 'perversions' were simply sexual activities that engaged with parts of the body or other material entities beyond those 'designed for sexual union', or that lingered at an intermediary stage in relation to the sexual object (1962: 16). And where in childhood, such perversions were seen as innate, their reappearance in adulthood is considered problematic only insofar as they signal incomplete libidinal substitution, or more specifically a failure of repression. Yet, it is repression – not perversion – that produces neurosis; as Freud puts it: 'neurosis is the negative of perversion' (1962: 104). In short, although genital heterosexuality may represent an ideal, it could be considered in the mode of a neurosis, while equally the openness of non-normative sexuality may take on a far more positive light.

There are strong grounds, then, for suggesting that it is time for a non-conventional appreciation of the Freudian oeuvre – and not only in the Lacanian mode. Although the poststructuralist approach which I broadly follow would always recommend a reading of Freud against himself, the seeds have already been sown in the original texts, albeit laying dormant. It is perhaps a matter of recovery as much as reinterpretation. The difficulty is that despite Freud's own ambivalence, the present-day psychoanalytic position remains unsympathetic to the appeal of polymorphous pleasures, seeing them as signalling emotional disturbance and abnormality, a regression to a pre-subjectival, pre-Oedipal state, or at very least as evidence of sexual immaturity. It is clear, that in the Freudian scenario, both psychic and ideological operations are at stake, and Freud insists that repression of the plethora of sexual objects in infancy is the necessary dimension of advanced organised societies in general. As he names it, the founding repression – and it is one that effectively instantiates sexual difference and the heterosexual matrix – is the prohibition of incest which is foreshadowed in the maternal–infant dyad. Yet it is not clear on a socio-cultural level why that supposedly universal law should also entail the abandonment of other less potentially disruptive elements of polymorphous pleasure. If at the psychic level, the castration complex and its resolution represent the mechanism by which desire for the mother is effectively covered over, then what is it that inhibits the expression of alternative forms of desire that speak not to prohibited relations between normatively embodied subjects but to differential forms of embodiment? Short of falling back on the principle that all forms of desire are mobilised by loss of the mother – which would not explain why heteronormative desire should alone escape censure – the appeal to psychic and social stability, and its implicit acceptance of regulatory controls, appears overplayed.

What this might seem to suggest is that were the parameters of appropriate adult sexuality less rigidly defined, if the polymorphous nature of sexuality were not suppressed, then the singular standard of genital heterosexuality would represent simply one possibility among a multiplicity of options. It would not be a matter of tolerating substitutes for the real thing, but of giving equal value to all; and in psychoanalytic terms, there would be a place for alternative forms of sexuality without a perceived loss of psychic well-being. We might, indeed, even expect an exuberance that marked the lessening of the anxiety generated by incomplete repression. The potential implications for a more positive model of sexuality specifically with regard to disability are considerable. Were the plasticity of sexuality to be acknowledged rather than repressed, then those whose bodily difference may quite literally preclude them to a greater or lesser extent from normative forms of sexual practice would neither be denied expression, nor devalued in their choices. The circulation of desire and the partial satisfactions of pleasure would be as much the unremarkable province of disabled people as they are for the ablebodied majority, rather than the site of overt disgust and shame. Could one not envisage that future enquiry into the differential specificities of embodiment and sexual preference would yield a way of giving positive value to the hitherto troubling reversals, the discontinuities and, yes, the perversions that might be seen as figuring a more open expression of sexuality? Put more precisely, the most optimistic reading is that the challenge to sexual repression would lift the burden of anxiety from normatively and anomalously embodied people alike. Certainly, then, there are some unexpected openings in the psychoanalytic canon, and in Freud in particular, but are they in fact enough to challenge the impasse of negativity associated with the conjunction of disability and sexuality?

In her attempt to the theorise feminine and more specifically lesbian desire against the grain of heteronormativity, Liz Grosz (1995) concludes that psychoanalysis cannot escape its conventional framing in masculinist sexuality, and in a similar way, I am inclined to doubt that the Freudian/Lacanian model could be sufficiently reconfigured to go beyond the assumption of normative embodiment.[15] Moreover, even were such a shift of emphasis possible, would the turn to polymorphous pleasures be enough to finally disturb the psychical link between desire and loss?[16] It is not that I think that a Freudian or Lacanian inspired analysis is off-track – quite the opposite – but that I have certain limiting doubts with regard to the specificity of the psychoanalytic approach. Is there then another way forward? By way of contrast to the analysis developed in this chapter, the decisive rejection of psychoanalysis enacted by Deleuze and Guattari (1984, 1987) enables a reconceptualisation of desire in terms of its productivity, and, by implication, of sexuality as a matter of flows, energies, and capacities that could always be otherwise. The dis-organised body that Lacan sees as the impossibility of stabilising a unified self who will become a subject in the Symbolic is rewritten precisely as the body-without-organs, the body that far from signalling separation and distinction persists only by making connections in the flux and flow of a desire that is without either a fixed aim or object. It is, among other things, a turn towards the positivity of intercorporeality, rather than an anxiety about its dangerous significations. Above all, desire itself – though no more fulfilled and closed down than in the rival model – is no longer figured in terms of lack, but is always directed outwards to establish ever renewed libidinal zones, that as Grosz writes, 'are continually in the process of being produced, renewed, transformed, through experimentation, practices, innovations, the accidents or contingencies of life itself' (1995: 199). Perhaps it is here that the conjunction between disability and sexuality could be more positively rethought.

I am reluctant, however, to pose the problematic in terms of an either/or choice between Lacan and Deleuze, and reaffirm that it may be possible to deploy multiple theoretical perspectives on the issues to hand without having to finally choose between them. Despite the evident antagonisms between a Lacanian and Deleuzian approach, I want to be able to hold them together and see the tensions as productive of a dynamic and intercursive understanding rather than as necessarily fatal to either. Where so little has been written directly of the interplay between disability and sexuality, it seems doubly important to keep open all the lines of enquiry. The trick is to maintain the theoretical rigour of each specific theory whilst at the same time conducting a critique that not only takes account of internal contradictions and impasses, but keeps a watchful and respectful eye on externally generated alternatives. If we are to gain both some understanding of what mobilises anxiety – albeit often in the form of fascination – in the normatively embodied majority, *and* a fuller expression of the sexual potential within disability, then clearly much work remains to be done. In Chapter 6, I shall explore the possibilities of a Deleuzian mode, which is finally beginning to be taken seriously within critical disability studies. Nonetheless, whether a singular or the cross-cutting approach that I am offering is preferred, what matters is that the whole area of sexuality and desire that has been largely unspoken should be opened up to innovative ways of thinking that go beyond the familiar and stultifying binaries of the cultural imaginary. Any new sexual discourse carries a certain disruptive danger, but it is a danger that all of us, however we are embodied, should be willing to embrace.

5
Transgressing the Law

> *After Freud, the task of Critical Legal Studies could well be that of listening to the suppressed.... Therein lies the possibility of justice*
> (Peter Goodrich 1992: 210)

Of all the discourses that may imbricate with the problematic of disability, the law is the most authoritative, seemingly offering not a speculative model like psychoanalysis that delves into psychic anxiety to explain governing norms, but an exhaustive and literal programmatic statement of the licit and illicit as it relates to bodies and behaviours. That is not to say, however, that the apparent externality of legal dicta is any more immune to the grasp of the socio-cultural imaginary than any other discourse. In this chapter, I shall both step back from the preceding close exploration of disability and sexuality to take up again the question of embodied subjectivity, and situate law firmly within an institutional nexus, which like the individual psyche, is fraught with anxiety and uncertainty. When legal theorist Peter Goodrich marks the necessity of listening to the suppressed, I hear an appeal to the significance of disavowed anomalous bodies. As I have suggested, the issue concerns not so much a conscious silencing of what may be troubling, as a form of repression that underlies any claim to rationality, impartiality, and consistency – precisely those attributes on which the authority of jurisprudence is founded. In incorporating its own imaginary, the law is fully imbricated with all the uncertainties, doubts, and anxieties that mobilise normative thinking in general and that create an unavoidable tension at the heart of what purports to be institutional neutrality. Where the hinterland of fantasy can be readily acknowledged as an element in public morality, and certainly in public moral panics – such as the witch-hunt for suspected paedophiles – it is less clear that *juridical* operations stand in the same relation to the unconscious. I shall not repeat the steps taken in the previous chapter with regard to psychic processes, but merely mark here that law is shot through with phantasy, the implications of which resonate at all levels of its operations. Bearing in mind that Derrida himself evokes Freud in his assertion that 'deconstruction is also the interminable drama of analysis' (1998: 29), my subsequent turn in this chapter to Foucault and especially to Derrida is not the methodological break that it might at first appear, but continues to indirectly pose the question of unconscious processes at work.

As previous chapters have made clear, all forms of differential embodiment and their associated behaviours – particularly in relation to sexuality, which is in any case a potent locus of disturbance to legal normativities – raise the question of what is legitimate and what is illegitimate. But that is already too limited, for, as Judith Butler (2002: 233) points out, the issue more properly concerns what is designated as *unintelligible*, in other words what is excessive to any binary of lawful and unlawful. In a context in which all practices of intimacy are subjected to an overbearing governmentality (a governmentality that is occluded, nonetheless, by the assumption of a public/private split that limits the juridical remit), the absence of specific law, or indeed social policy as the local implementation of law, cannot be read as simple disinterest. Rather it is the point at which 'plain meaning' gives way to what Goodrich sees as a state of legal delirium. In his book *Oedipus Lex: Psychoanalysis, History and Law*, Goodrich (1995) argues that a genealogical approach to the history of law would be an enquiry into an institutional imaginary that is founded on exclusion. What is claimed is that despite the supposed impersonality of jurisprudence, in the moment of speaking in the Symbolic, the judge must disavow his other – whom Goodrich names as the feminine, the one who is now lost to the subject, but is constitutive of him. This is a move that has strong parallels in feminist theory in the work of Luce Irigaray, but I should like to leave aside sexual difference (though the relation between masculine and feminine is never insignificant) and extend the argument – as I did in the last chapter – to corporeal difference. My own contention is that the form of juridical (self-) effacement theorised by Goodrich is strongly paralleled in the disavowal of the disorganised body of disability. That should not imply that the insights of a psychoanalytic approach can be literalised in terms of the juridical or socio-cultural order, but that the contradictions and discontinuities, and the attendant anxieties which mark everyday jurisprudence and social policy, find their roots in unresolved and irresolvable psychic tensions, and most particularly in the impossible structure of desire, in its full Lacanian sense.

Tempting though it is to pursue that thought further, my interests in this chapter are more constrained, and I shall leave it simply as a provocative suggestion. Nonetheless, the complexity of desire, in any case, encompasses both sexuality and differentially embodied subjectivity which are my themes, and is deeply implicated in the mobilisation of disavowal. With regard to the former, the point is that in seeking to exclude from its parameters anything that might count as less than 'normal' heterosexuality, the law must either actively bar the other and thus afford it at least negative recognition, or disavow the troublesome other altogether. As with social policy, my reading of legal process engages with both forms of silencing, but it is the latter that leads directly to the problematic not only of the fantasy structure of the cultural imaginary, but of the phantasmatic structure of the unconscious. In the cultural imaginary fantasy operates to both construct and resist the operations of law, while in the unconscious phantasy is that which organises the subject itself. The coherence of the legal imaginary is then always at risk, perhaps most dramatically in relation to sexuality, precisely because the wider cultural imaginary is both represented by and saturated with the institutions of heteronormativity. Indeed, as Lauren Berlant and Michael Warner point out, heterosexuality is the 'space of pure citizenship' (1998: 549). To reiterate, I am not suggesting that disabled people have any more propensity for a non-heterosexual organisation of sex and sexuality than other group, but what is clear is that their prevalent forms of sexuality are less likely to fit the *normative* paradigms of the phallic masculinity that incorporates both male and female actors. In consequence the conjunction of sexuality and disability is a prime site of anxiety insofar as it encompasses two forms of non-normativity. The terms of the cultural imaginary, and more specifically the legal imaginary, are put under threat, resulting not simply in exclusion but in a form of disavowal.

The striking paradox that I intend to uncover here is that if law were indeed the rational, objective, and programmatic operation that it claims to be, there would be no need for disavowal, either of supposedly threatening others, or of its own troubled foundation in originary violence.[1] In reality it is inconceivable that any discourse can escape the tissue of connections from which it emerges, but the language of jurisprudence in particular – which Goodrich (1987) bluntly describes as 'profoundly alien ... archaic, obscure, professionalised and impenetrable' (cited in Davis 2002: 120) – claims authority precisely by covering over the messy and uncertain contexts that constitute its condition of possibility. Discourse is always already imbricated with its others, shot through with impasses, metaphor, contradictions and disavowals, and those aspects, as both psychoanalysis and deconstruction have shown, are not states of failure but the very condition of discourse at all. The task, then, of cultural theory in general, and for my own purposes critical disability theory in particular, is to critically interrogate 'the pathologies of modernity' (Barron 2000: 303) by opening up the questions that are left unexamined by the intellectual parameters of the postEnlightenment. This entails both a move to uncover what has been suppressed and repressed, and as such is deeply disruptive to the cultural imaginary, and a related commitment to explicate how the very process of specifying material, disciplinary, categorical, and ethical boundaries carries within it the paradoxical possibility of deconstructing – taking apart – those very same distinctions. Insofar as conventional jurisprudence relies on clear and distinct ideas, it is paradigmatic of Enlightenment discourse, and for that very reason susceptible to radical reconfiguration on both ontological and epistemological grounds. In radically deconstructing – in their varying ways – the institution of law, Foucault and Derrida are theorists who carry forward the project of contesting the normative certainties that seem to sustain the assignment of disability to a devalued and disavowed discourse.

At a fundamental level, the law must always be implicitly concerned with embodied subjectivity, but the body that it recognises is a fabrication that stands against the lack of stability and unity that is the condition of embodiment. Nonetheless, to be acknowledged as such a subject before the law is the basis for claims to the justice, legal rights, and protection from which some others are excluded. At the same time, such recognition is also the foundation for a thoroughgoing governance that undermines the very attribution of autonomous agency that grounds modernist subjectivity in the first place. What emerges in terms of critical theory, then, is the need to attend both to the structure and procedures of modernist juridical practice, and to some more abstract questions regarding the constitution of the subject, and the nature of justice itself. In the interface between law and disability, the attribution of pathology, as Barron suggests, might adhere more appropriately to the juridical than to its conventional target of the disabled body. Once again I turn first to Foucault who gives an extraordinary analysis of the historical relation between bodily anomaly and law that is rich in insights into the inherently transgressive nature of disability, before taking up the better-known Derridean deconstruction of law that serves to make plain why disability poses such a troubling challenge to the structures of justice. The deconstruction of normativities, which is strongly but differentially forged in the work of both, continues to theoretically ground transgression not in the self-regarding play of cultural rebellion, but in a deadly struggle against what manifests, above all, as the force of law. What is at stake here is the corporealisation of transgression in anomalies that, though they appear arbitrary, and sometimes monstrous, speak to the undecidability of all forms of embodiment. Transgression is not a feared or desired outcome of difference, but exposed as the *retrospective* default position.

In his 1974–1975 lecture series published as *Abnormal* (2003a), Foucault makes some startling, yet thematically consistent, claims about what constitutes the monstrous. It is not the otherness of morphological abnormality in and of itself that is his focus, but the offence it offers to the law. As he understands it, the transgression of bodily form matters primarily in view of its impact on a series of established laws, customs and regulations setting out what is proper to a particular moment in socio-cultural history. This counter framing of the monster in a quasi-juridical context – which goes well beyond any offence to the so-called natural order – enables Foucault to name conjoined twins as the privileged signifier of the early modern meaning of the monstrous, precisely because such figures are uncontainable and undecidable within civil, canon, or religious law. Although Foucault goes on to outline some strategies of normalisation in response to anomalous bodies and minds, and to chart the collapse of the 'great' monster as a figure of difference, the problematic of the insufficiency of law remains. In quite another register of analysis, Derrida too unsettles the law, not least by posing its imbrication with what he names as the 'monstrous *arrivant*' (Derrida 1995). For both writers, the law is never impartial but always caught up with the strategies of power and a discursive violence that seeks to grasp and domesticate the troublesome other. Yet where Foucault sees resistance inherent within the productive power of normativity and governmentality, for Derrida, the law must be transposed from that sphere – where it functions as *droit* – and rethought in relation to an impossible justice. When Derrida speaks of justice, he refers not to the here-and-now application of law through the agency of the police and the courts, but to that which is always yet to come.

What, then, can the critical intersection between these two highly individual notions of the law offer to an exploration of the substantive ground of certain anomalous bodies? My purpose here is to trace the import of the monstrous in general, and the uncertain parameters of

disability in particular, in relation to Foucault's and Derrida's speculations on the discursive meaning and significance of respectively law and
justice. Although Foucault himself makes a distinction between monstrosity and disability on the grounds that only the latter is encompassed
within legal frameworks, it is not one that stands up to scrutiny, and
indeed his own exposition of 'monstrous' legal cases involving people
with intellectual disabilities demonstrates the historical instability of the
terms. In any case, I shall take up his own reading of conjoined twins,
not simply as a signifier of the monstrous (the early modern nomination), but as the limit case of the disabled body (in its contemporary
conception), to explore why – contrary to Foucault – the everyday law
can never resolve the problematic of disability. As the increasing deployment of queer theory in contemporary disability studies shows, the
possibility of a radically transgressive form of embodiment, which – in
relation not only to limit cases but to more familiar corporeal anomaly –
is finally beyond the disciplinary reach of normative paradigms, remains
open. The question is whether the appeal to what is often seen as highly
abstract theory can yield new ways of thinking otherwise about the
materiality of those figures of anomalous embodiment that frustrate
social, cultural, and legal normativities. Moreover, if disability in its
many forms always in some way transgresses the law, then how is it to
be thought ethically? What will ground response and responsibility to
the other who exceeds the confines of regulation, who cannot be held
within a category? These are the questions that most occupy Derrida,
and I shall suggest finally that he finds a way forward in the notion of a
non-provisional hospitality.

Turning to Foucault's early work, and particularly to *Discipline and
Punish* (1977b), it is clear that the major trajectory is concerned with
what he sees as the radical change from a form of society in which the
figure of the Sovereign embodied the law and authorised its repressive
and always potentially violent power, to one in which power was dispersed into multiple microphysical and disciplinary channels that are
directed towards the bodies of all, regardless of their legal standing, or
compliance with external markers of the licit and illicit. The emergence
of biopower as the instrument of regulation, at the level of both popula-
tions and individuals, marks for Foucault the modern form of power in
which there is no undisputed central agentic force – neither as Sovereign
nor State – that wields control, but rather a multiplicity of decentred
disciplinary technologies that both constitute and categorise embodied
subjects according to an inescapable system of normativities. It is not
that authoritative physical force entirely disappears, but that discursive

power is more effective when it pacifies by way of procedures that the emergent subject may comply with through apparent choice. As Judith Butler puts it: 'the law is not literally internalised but incorporated, with the consequence that bodies are produced which signify that law on and through the body' (1990: 134–5). The principle organising binary is no longer legal and illegal, but normal and abnormal; and the target is not simply forms of external action, but the totality of embodied being as it manifests both in the inwardness of psychic processes and corporeal self-awareness, and in the exteriority of comportment and behaviour. In short, it involves all the terms of self-identity.

In any superficial reading of Foucault, what appears to happen, then, is a historically situated shift in which the privilege of law is displaced and superseded by non-juridical modes of power, principally in the form of biopower. But that would be a mistaken assumption, for at very least, as Wendy Brown and Janet Halley comment, Foucault himself

> has taught us to be alert to the imbrication of juridical and disciplinary discourses, an imbrication that can range from the overt to the very subtle. Indeed, law has a penchant for hiding itself in background rules so minute that they facilitate or activate regulatory regimes that seem immune from legalistic effects.
>
> (Brown and Halley 2002: 11)

This is a more sophisticated understanding that goes beyond law's easily recognisable regulatory function to implicitly contest its status as primarily either protective or repressive. As with all forms of disciplinary processes, the question of right or wrong is suspended in favour of an understanding of power as productive. The subject no longer stands before the law awaiting judgement, but is constituted by the subject effects of its capillary circulation. And those effects are sustained not because they signify a static entity, but precisely because the embodied subject is caught up in a system of *reiterative* norms, including those that appear to bestow rights. What is perhaps more interesting about Brown and Halley's analysis, however, is that it opens up the possibility of extending that persistently pervasive and interwoven notion of law back to the premodern and early modern period in which power was invested directly in, and violently enforced by, the head of state. If in the era of modernity, 'the law' exceeds, as they put it, 'the figure ... of the prohibiting, death-wielding sovereign, and has incorporated the managerial, normativizing, regularizing, biopoweristic forms that [Foucault] proposed were distinguishable from the juridical form, even if historically entwined with it' (2002: 13), then we must suppose that it has always exceeded its most overt form. If the law today 'is capable of intensely intimate effects' (13), then it has been so in the past. And although Foucault resists the move I have described, this is precisely what his reflections on the problematic of the abnormal seem to reveal. Then, as now, some bodies are deemed so extraordinary that they stretch law to the limits of its intelligibility, and yet that law – infinitely adaptive – never ceases to extend its domesticating grasp.

Before exploring further that conundrum for contemporary law, I want to return to Foucault's lectures on the abnormal. His thesis is that the domain of abnormality – which by the nineteenth century has been brought together under a 'system of regularities' – is constituted by three elements distinguished and defined during a previous period. Of these the first and earliest – the human monster – is my immediate concern. Drawing on a series of archival texts, Foucault outlines how the monster stood for 'the transgression of natural limits, the transgression of classifications, of the table, and of the law of the table' (2003a: 63). In its exemplary hybrid form, the monster posed a series of highly disturbing and practically irresolvable questions that pushed the law to the point of rupture in each of its strongholds: in civil society, religion, and the natural order. Was the offspring of human/animal fornication truly human? Should such a bestial child be baptised? Should a conjoined twin be executed for a singular crime? Whom could a bisexed monster marry? Did the monster have inheritance rights? These and similarly troubling concerns were widely debated, not for the betterment of those with radically anomalous forms of embodiment, but because they challenged the stability and authority of law in its various guises. As Foucault remarks, the hybrid is named as monstrous 'only because it is also a legal labyrinth, a violation of and an obstacle to the law, both transgression and undecidability at the level of the law. In the eighteenth century the monster is a juridico-natural complex' (2003a: 65).[2] For two centuries at least, conjoined twinning was the most highly remarked form of the monstrous, gradually losing prominence to the figure of the hermaphrodite; and it is perhaps no surprise that the questions that so disturbed the authorities of the early modern and classical age should find their modern counterpart in the twenty-first-century concerns with the legality and morality of separating conjoined twins, or with the legal identity of transsexual and intersexed individuals. Where Foucault's concern is to trace a move from the juridico-natural by way of the juridico-moral to the monstrous nature of criminal deviance as such, it should not be supposed that those earlier frameworks exhaust their significance in the historical

understanding of criminality. Faced with the profound unintelligibil-
ity of anomalous embodiment, what has changed in contemporary
times is not the problematic itself, but the way in which it is managed. As Foucault (2003a) notes, the technology of an essentially violent
and repressive power is superseded by a positive power that creates the
very subjects it manages. And the law is implicated equally in both
strategies.

What is at issue in the historical response to the monstrous is not,
however, so much an originary campaign of violence against an alien
and unknowable other, as a *failure* of less repressive strategies. If, as I
have suggested, law always exceeds its prohibitive power, that is not to
say that it does not resort to forceful rejection in the face of radical transgressivity. Nonetheless, what the archival texts show – including those
on which Foucault relies[3] – is that the questions posed are precisely
those that would attempt to recuperate the anomalous or impaired
body within some regulatory mechanisms. When, for example, the
seventeenth-century canonist, Alphonso a Carranza lays down the law
of baptism appropriate to human monsters, or Ambroise Paré lists in
1573 the 13 causes of monstrosity, or even when Augustine in his
early fifth-century text, *City of God*, speculates that monsters will be
restored to perfect human form at the Resurrection, there is already
some normalising impulse in play.[4] Foucault too recognises this possibility when he remarks elsewhere: 'In general terms... the interdiction,
the refusal, the prohibition, far from being essential forms of power,
are only its limits, power in its frustrated or extreme forms' (1988b:
118). Whilst the encounter between those with hybrid form and the
conventional legal norm of singular self-identity is likely to result in
confusion and ambivalence at least, an understanding of the dispersed
domain of the juridico-political demonstrates that there is rarely an
absolute exclusion – or escape – at work. Legal and disciplinary power
work together. Despite, moreover, the ostensibly uncomplicated contrast that Foucault sometimes draws between old and new models of
power, and indeed between the monstrous and disability, the reality
is more complex. Where his striking example of the change from the
exclusion of lepers to the inclusion and management of plague victims
(2003a: 44–5) is intended to illustrate the shift from negative to positive power, Foucault acknowledges that the two continue to overlap,
and that the latter model 'was not established but reactivated' (44). In
any case in his own account of disability, Henri-Jacques Stiker, who is
broadly sympathetic to Foucault's history of *mentalités*, claims that even
in the Middle Ages, people with bodily impairments were seen both as a

category to be offered charity – like the poor and sick – and as strangers who should be exorcised from the community (Stiker 1999: 69). Certainly the centre of gravity may change as Foucault asserts, but there has never been a singular model in operation.

It is with those provisos in mind that the relationship between the law and the monstrous can better be understood. Like Foucault, Stiker is reluctant to conflate physical disability with monstrosity, although he concedes that 'the *notion* of the monster is necessarily related to that of disability' (1999: 72). The difficulty arises only because, for him, the former term signals towards the so-called monstrous races which in pre-modern times were supposed to inhabit the outer margins of the civilised world, effectively therefore beyond the reach of civil and canon law, if not natural law. It is precisely the contrast with these races that interests Stiker insofar as, '(t)he disabled, the "monsters" immanent in our society and not on its borders, heighten our fears, because *they are already there*' (1999: 70). In Stiker's view, Foucault – despite his impressive work on madness and the era of the Great Confinement – totally neglects to explore physical disability. I should prefer to say rather that Foucault brackets off the more severe cases of congenital disability as a form of the monstrous, pays little direct attention to minor anomalies, and shows no interest in acquired disability. Nonetheless, despite such apparent and putatively unjustified divisions, he is clearly aware of the irreducible links:

> [The monster] is the magnifying model, the form of every possible little irregularity exhibited by the games of nature....the monster is the major model of every little deviation. It is the principle of intelligibility of all the forms that circulate as the small change of abnormality.
>
> (2003a: 56)

Indeed, he says something very similar to Stiker: namely that the monstrous is a notion *within* the law, yet that it is its challenge to the limits of the law that creates anxiety. What is perhaps different is the degree of ambiguity that Foucault insists on, and it is an ambiguity that is most clearly developed with respect to those forms of embodiment that lie both at the very edge of the possible, and at the point at which law is overturned. While it is difficult, on the evidence cited here and elsewhere, to agree with Foucault that the monster 'violates the law while leaving it with nothing to say....the monster is a breach of the law that automatically stands outside the law' (2003a: 56), his observation, that '[p]aradoxically the monster is a principle of intelligibility in spite of its limit position as both the impossible and the forbidden' (57), is rich in significance. It is here surely that the potentially transgressive nature of every deformity or physical anomaly emerges, and the reason why disability stands in a relation to the law that can never finally be secured.

Throughout the course of the lecture series on the abnormal, the point is made that abnormality is the common denominator across the discursive chain that slides between unusual forms of embodiment, mental instability, deviant behaviour (particularly with regard to sexuality), and criminality. The insult of the monstrous in its grand form is gradually reduced to the everyday, and as Foucault puts it: 'The recurring problem of the nineteenth century is that of discovering the core of monstrosity hidden behind little abnormalities, deviances, and irregularities' (2003a: 56). The ambiguities do not disappear, but they are 'toned down and muffled' to become implanted in the whole problematic of the abnormal around which judicial, medical, and later psychiatric technologies revolve. It is in such a context that disability itself is to remain suspect, regardless of the disciplinary strategies directed against it, and the degree to which it is effectively covered over in modern-day civil society. It is Stiker's view that the normalisation that disability undergoes – particularly in the twentieth century – through strategies such as rehabilitation and formal legal equality can be characterised as constituting the erasure of the difference of disabled people. As he remarks: 'the will to assimilate, to trivialize by an intense circumscription and treatment, cannot be challenged' (1999:192). But that, I think, posits a too easy view of the all-encompassing success of discursive power, and fails to capture the cracks in the facade of assimilation and homogeneity. As Foucault has always made clear, the operation of power repeatedly generates its own refusals, its own counter-discourses, but there is more at stake even than the performativity of resistance. It is, rather, a question of the haunting sense of a remaining undecidability that is never entirely settled or resolved by the technologies of power that shape the notion of disability. To the extent that disability remains caught up in the signifying chain around abnormality – a link that is continually highlighted even as the juridico-social seeks to expunge it, it continues to contest final identification. And as Foucault puts it: 'At the forefront of [its] ambiguities is the never wholly mastered interplay between the exception of nature and the breach of the law' (2003a: 324). Law, in other words, is never sufficient to arrest the disruptive possibility, indeed certainty, of an excess inherent in all forms of disability. Transgression is unavoidable. Yet the point is not that disability is unique in its capacity to evade the normative grasp, for all corporeality is leaky and unpredictable (Shildrick 1997); rather, it is indeed the 'magnifying model' of the 'normal' body.

Before turning to a closer analysis of that transgression, I want first to set out in more detail the problematic operation of the techniques of normalisation directed to the disabled body. Just as Foucault (1977b, 1979a, 1980) would imply, and as Stiker traces with more specific application, the disciplinary and regulatory processes in play both constitute and attempt to contain the disabled body within highly determined parameters.[5] In the twenty-first century, biomedicine, social policy, cultural discourse, and psychiatry all play a significant part in the policing of boundaries, but it is changes within the law – in its narrow sense – that have been most highly concerned with the definitional work that marks out disability as a discrete category. For the most part, legislation in support of people with disabilities has taken as its baseline some notion of justice, understood either as the formal equality of opportunity, or as a distributive model which compensates those unable – by reason of their impairment – to take advantage of the available options. Both the UK Disability Discrimination Act (1996) and the Americans with Disabilities Act (1990) offer protection under the law that rests broadly on the social constructivist model, which posits disablement as arising from inadequate or discriminatory social responses to bodily impairment, rather than from the impairment itself. Additional legislation under social security provisions is more closely derived from the medical model of disability, which, in seeing the problem as residing in the individual's own body, is more concerned with appropriate care and compensation than with accommodation. Although the most forceful impulse of disability activism and advocacy has appealed to the social model in the search for justice, both approaches are engaged with rights discourse, particularly in the United States where the notion has a potent history. In parallel with my previous remarks about governmentality, the problem is that despite the resultant material benefits and enhanced social status for many disabled people, the appeal to legally situated notions of justice and equality throws up many irresolvable concerns. Any serious account of rights will need to acknowledge the multiple complexities that inform the differences between positive and negative rights and their relation to diverse forms of justice; the issue of conflicting rights; the problem of differential access to law; and even the question of whether rights should be preferred to alternative values. But the greater problem surely lies in the continuing failure to recognise either the disciplinary effects, or the irreducible incompletion and instability, of any such discourse.

Turning to the first issue with regard to the regulatory dimension of legal discourse, Foucault leads the way with his unequivocal recognition that 'the law operates more and more as a norm, and that the judicial institution is increasingly incorporated into a continuum of apparatuses (medical, administrative, and so on) whose functions are for the most part regulatory' (1979a: 144). As a consequence, a by now familiar concern arises that normalisation, which demands homogeneity – named in the juridical domain as equality before the law – inherently risks effacing difference, at the very moment of appearing to recognise it. For marginalised groups, including people identified or identifying as disabled, the urgency of mitigating the gross effects of discrimination tends to occlude the dangers of a move from exclusion to putative integration. Writing of the difference of disability, Stiker notes: 'social agencies have to intervene, to provide relief...but first of all to rid us of it, in a real or symbolic way' (1999: 7). Where once the delimiting strategies of institutionalisation hid disabled people from social sight, the twentieth-century emphasis on rehabilitation within the community marks what Stiker sees as integration to the point of oblivion, achieved not least through legal statutes. Echoing Foucault, he writes: 'To legislate is to impose a norm, a code of the universalist type on phenomena and practices which until then had been left on their own, to their diversity, empiricism, even their anarchy' (1999: 134). Although in the very act of designation, the disabled body is produced in *multiple* ways – as blind, as mobility-impaired, as congenitally deformed, as accidentally or temporarily inoperative – each with its own specificities and norms, the transgressive possibilities that have marked the monstrous seem reduced to conformity and docility. In short, the specific extension of rights accruing to disability cannot be counted an unproblematic good, but as an intensification of the disciplinary grasp of biopower.[6]

What is at issue, as Wendy Brown notes, is that 'rights are never deployed "freely," but always within a discursive, hence normative context' (2002: 422), which is of course the very context in which an oppressed identity category – in this case, disability – has been constructed, assigned, and reiterated as such. Besides, it is not the wider society that is constrained to change but always the individual marked by difference. In order to attain the full range of civil rights and benefits, that individual must remake herself as a legal subject – in both the active and the passive Foucauldian sense – who is consistent with the requirements of normativity. Where that is not possible, she may be accorded what are called 'special' rights that serve to decisively denote her failure to achieve the privileged standards, and that position her difference as an inadequacy in a hierarchical binary.[7] Each time the law comes to the assistance of a disabled person with relief or protection, she is reinterpellated precisely within that devalued identity, and thus marked for further regulation. To call on the law *as* disabled is scarcely a challenge to the normative standards of ablebodiedness that tacitly underlie the liberal humanist notion of a legal subject – that is, one who exercises independent agency – but serves rather to unavoidably consolidate the power of the system that constitutes and sustains such binaries in the first place. The options, then, appear to be between some form of assimilation that in operation covers over, rather than addresses, difference, and an attention that fixes difference into a damaging hierarchy. The outcome for disabled people is perhaps similar to that which has been endured historically by women in societies dominated by masculinist norms.[8] Of that familiar territory, Brown writes:

> The paradox, then, is that rights that entail some specification of our suffering, injury, or inequality lock us into the identity defined by our subordination, and rights that eschew this specificity not only sustain the invisibility of our subordination but potentially even enhance it.
> (2002: 423)

And, as she notes, disadvantaged people cannot not want the rights and recognition that the law both promises and denies, thus further ensnaring themselves in the categories of the dominated. It is a paradox that sets the stage for the move from the law as normative and regulatory to law – in the mode of justice – as finally undecidable.

Before taking that step, however, which will shift the focus from Foucault to Derrida, I want to briefly return to the problematic of conjoined twins, whose existence in the contemporary world can be as ungraspable and challenging as it was for the early modern period. I have touched elsewhere (Shildrick 2005b) on the strained relation between the laws of sexuality and disability in general, and more particularly on the peculiarly unremarked status of conjoined twins – with regard to consent, marriage, adultery, and so on – that suggests an excess that is beyond the scope of judicial discourse. Here, however, my concern is with the failure of the law as applied. For all conjoined births, the biomedical and social assumption is that separation – whether swift or delayed – is in the best interests of the infants and children involved.[9] The usual conceptualisation of the 'problem' as involving the need to free two putatively singular individuals from the monstrous embrace of a conjoined body speaks eloquently to our society's privileging of autonomous selfhood as the only right and proper form of human being (Nedelsky 1989; Shildrick 1997). And separation surgery – as with the conceptually related procedures of transsexual or intersex surgery that seek to stabilise gender by way of fixing biological sex – is perhaps the most extreme instance of the regulatory technologies that are deployed in the interests of normativity. The dictum that power is productive even as it intervenes is clearly illustrated in the transgressive body by the manner in which the corporeal cut is not destructive as such but precisely a means of subjectivisation. The ideal outcome – that two self-sufficient persons should replace the undifferentiated one – enacts precisely Foucault's claim that the subject is brought into being only to the extent that the body is subjected to those processes of normal-isation that both individualise and yet impose uniformity. But what happens when the law is called on to adjudicate on such anomalous embodiment *prior* to the emergence of the individuated subject? How are entitlements and rights assigned in the absence of a singular legal subject, and in the face of a radical uncertainty about the parameters of sovereignty over the body itself?

The case of the Maltese twins known as Jodie and Mary, who were not simply conjoined but fundamentally concorporate, came before the British courts around just such issues (*Re A* 2001 *[Conjoined Twins: Medical Treatment]*). Soon after birth it had become apparent that the twins could not both individually survive separation surgery, and equally that in remaining as a single entity – Jodie–Mary – 'both' would die.[10] In part, the question facing the court, then, was whether sep-aration could be legally authorised even though it would result in the inevitable death of the putatively weaker twin, and whether it *should* be authorised given a conflict of interests between the parents – who favoured no action – and the attending physicians – who urged intervention on behalf of Jodie, the probable survivor. Faced with the dilemma of indeterminate corporeality and a doctrine favouring the 'best interests' of the twins themselves, the court's consideration of the parental/professional divergence with regard to consent to treat was, however, of secondary importance, at least in the ethical sense.[11] And indeed, as the appellate hearing progressed it became apparent that the major issue that the court was attempting to resolve was how to distin-guish between the interests of Jodie as opposed to those of Mary, prior to their clinical instantiation as singular subjects. In other words, it was not so much the undecidability as such of the twins' concorporation that demanded a judicial decision, but rather that legal recognition of radical otherness is so highly constrained that the law's operations with regard to entitlement can function only insofar as it can identify the other within normative parameters. In short, it is only *the individual* who can be interpellated by the law. For Jodie–Mary that meant paradoxically that the decision about the legality of separation was conducted *as if* they were already separate beings, albeit ones engaged in a life and death dispute over limited bodily resources. Thus, Ward L. J., in referring to the parents' wish that the twins should remain untreated, could suggest that 'in their natural repugnance of the idea of killing Mary ... (they) fail to recognise their conflicting duty to save Jodie' (*Re A:* 53). In a similar way, though the opinions of the Appeal Court inspired plentiful critique, little of it questioned the certainty of that supposed division. In an otherwise nuanced analysis, Helen Watt, for example, asserts of the twins that '(i)t was clear ... that there were two separate systems of self-organisation, despite some overlap in the parts controlled', adding weakly 'I am referring ... to self-organisation from two separate sources, however this was achieved' (2001: 238).

For all that the courts and much of the legal and bioethical commentary on the decisions shows low awareness that the difference of concorporation might at very least problematise not simply the privileging, but the very concept of the autonomous sovereign subject,[12] it is clear that Jodie–Mary's embodiment cannot be normalised without both discursive and material violence. The very attribution of separate forenames is simply a linguistic trick that attempts to occlude the indeterminacy of their status as either one or two, while the enforced cut instantiates what has already been conceptually decided. Despite being projective at best, and clearly not identical with the newborn, it is the post-separation persons who are the object of the debate. In effect, the Jodie–Mary infant is unrecognised as a legal subject, and her place usurped by the Jodie and Mary who, ideally, *will* come to exist as individuals. Without that sleight of hand, it is clear that as a normative and normalising structure in the Foucauldian sense, the law is unable to adjudicate on the transgressive body itself. The whole structure – either of deontological rights and duties or of utilitarian measures of harms and benefits – deployed in the question of conflicting interests cannot address the undecidability of radically anomalous embodiment. And although conjoined twins like Jodie–Mary might be seen as exceptional, my claim is that in premising its strategies on the assumption of autonomy and self-sufficiency, liberal jurisprudence will always falter in the face of those whose embodiment signals a confusion of corporeal boundaries and/or a necessary connection and interdependency. Where the normalising power of law is sufficient to regularise and contain within a category many forms of disability, more challengingly transgressive corporeality – although even the everyday potential of slippage between mother and foetus inherent in pregnancy is sometimes problematic – may resist its disciplinary drive altogether, but only at the cost of forfeiting the positive value of legal protection.

How then should the transgressive be conceptualised? Clearly the accommodation of some forms of embodiment within the legal structure remains impossible insofar as those forms do not simply escape its instrumental details, but radically disturb the very framework under which the juridical operates. It is a disturbance, it seems to me, that goes beyond the Foucauldian cycle of power and resistance that characterises the processes of subjectivisation. In any case, if recognition under the law is a double-edged sword that implicitly signifies dominance and normalisation, then it is time to consider whether the transgressive might be resignified as a productive positivity. Where the subject grasped by the law is centred, stable, and atomistic, and must necessarily exclude its others, perhaps the Derridean notion of a dispersed and undecidable subjectivity shows an alternative route. The conventional mode of autonomous self-sufficiency might give way to an embrace of difference that is celebrated precisely in its uncertainty, its fluidity, and its interconnections. Many forms of disability are overtly implicated here, but it is instructive to go further to reflect that the norms of self-sufficiency and bodily integrity are in any case illusions that cover over the impossibility of a fully autonomous individuality, and indeed the inescapable connective vunerability, of any one of us (Shildrick 2002). Similarly, although the law appears as a largely inescapable and compulsory force, it is as well to remember, as Ziarek remarks, that

> (it) is marked by the 'infelicities' and the infidelities characteristic of performative utterances. The repetition of acts understood as the citation of the law stabilizes the form of the law, and, at the same time, produces a 'dissonance' and inconsistency within it.
> (1997: 128)[13]

Such an infidelity and dissonance occurs precisely, I would suggest, in the unremarked Court of Appeal slide from the being in the body of the infant Jodie–Mary to the instantiation of Jodie and Mary as distinct legal subjects. Neither embodiment or subjectivity, nor the law itself is as closed and secure as it might seem.

For Derrida that same notion of undecidability is at the heart of what he understands as the impossibility of justice, not simply *in* the present, but as an active presence as such. As he puts it, justice is, and will be, always yet to come, the *a-venir* that acts as a corrective to the operations of the law. In his most extensive explication and deconstruction of the law, 'Force of Law' (1990), Derrida shows that it is precisely the inevitable failure of the legal system to successfully bring everything and everyone under its remit that forces the realisation that the ethical necessity of responsive and responsible decision-making must function at the limit without guidelines, norms, and precedents.[14] This is a familiar theme in Derrida's reflections on ethics, applied here to the delivery of what is supposedly just by the operation of law. Contrary to the authoritative voice of the judiciary, Derrida insists – and he could be speaking precisely of Jodie–Mary – that

> (l)aw (*droit*) is not justice. Law is the element of calculation, and it is just that there be law, but justice is incalculable, it requires us to calculate with the incalculable; and aporetic experiences are the experiences, as improbable as they are necessary, of justice, that is to say of moments in which the decision between just and unjust is never insured by a rule.
>
> (1990: 947)

It is not of course that the everyday conventions always falter, but that what they deliver always falls short. Yet far from having a paralysing effect on judgement, the ab-sence of justice is the very condition under which juridical discourse makes sense at all. The paradox of the *impossibility* of justice – its coming infinitely deferred – and the *need* for justice now exposes the problematic of law to recasting, transformation, and responsibility to and for the alterity of the other. As John Caputo puts it in his celebration of deconstruction: 'existing legal orders are exposed to continual correction, revision, and alteration in the light of the claim laid upon them by alterity, that is, by what is beyond that code, by what is silenced or excluded by that code' (2000: 132). What the deconstruction of law demands is not the demise of legal structures, but an openness to the transgressive claims of the other. It cannot be said that we *have been* just, only that we strive for justice.

To bring the irreducible other under the remit of the law is always to effect a certain violence against her that reduces her difference to the parameters of the selfsame. The legal recognition of the other is both an urgent need – as Derrida notes: 'Nothing seems *less* outdated to me than the classical emancipatory ideal' (1990: 971) – and a move that entails dominance, normalisation and domestication. The claims of those others who are not like us, whether literally or metaphorically, are deeply unsettling to the extent that, as Derrida puts it, 'as soon as one perceives a monster ... one begins to domesticate it, one begins ... to compare it to the norms, to analyze it, consequently to master whatever could be terrifying in this figure' (1995: 386). To that extent, Derrida's explication coincides with Foucault's understanding of the transgressive as either wholly outside, or else as recuperated to normativity within the law, albeit with the possibility of resistance. Where Derrida goes further, however, is in his suggestion that transgressivity is inherent in the juridical structure from the beginning. Although the infelicities at the margins may be the more evident, *no* degree of normative practice or compliance is sufficient to finally evade the uncertainty and provisionality of all legal process. The law is destabilised not only by the external other but by alterity within. And it is here that the impasse implicit in the legal order becomes apparent: where the demand to treat all equally – to impose some form of identification that enables universal application of law – runs counter to the need to treat each in her singularity and difference. As Drucilla Cornell writes: 'For Derrida, the most fundamental aporia of justice is just that: justice must be singular and yet justice as law always implies a general form ... if it is not to be blatantly unjust' (1991: 113). In conventional thought, if the law is just, then it must necessarily re-present the alterity that is alien to the commonality assumed by the concept of legal equality; yet *singular* justice must exceed every judicial structure. In Derrida's terms, justice is finally aligned with the Levinasian notion of response to the incommensurable other, whose need and demand is thus impossible to satisfy.

But can such a paradoxical abstraction – such undecidability – mobilise any significant reworking of the substantive relation between law and the figures of difference who come before it every day? Again, it needs to be said that the incommensurability of the other is not peculiar to those at the margins, but epitomises each and every encounter to the extent that law is never fully adequate. Nonetheless, there are particular transgressions of the normative, such as greater or lesser corporeal anomalies, that attract especial injustice. Moreover, Derrida himself is clear that although the critique encapsulated in 'Force of Law' concerns 'juridico-politicization on the grand geo-political scale' (1990: 973), it must also encompass other seemingly secondary and more material concerns, which he goes on to partially list. Among these are several current bioethical issues consonant not only with the legal response to conjoined twins for example, but with the wider problematic of disability.[15] As Derrida implies, the claim of the other on each of us is concrete, and the demand is for a response and responsibility that does not defer, but engages with alterity now. His project, then, is actively committed to the ethical, not as yet another set of normative standards, but as a projective exercise that is open to the risk of undecidability and uncertainty. To think beyond law or programmatic morality, to think outside the familiarity and comfort of rules and principles, to resist the urge to reduce the other to an object and category of knowledge, is the real test of responsibility. As Caputo puts it:

> Maybe what is coming is nothing as simple and unambiguous as an hermaphrodite or an androgyne, but something undecidably misce-genated, something that has not happened yet, something singular, something possible, something impossible, something unimaginable and innumerable.
>
> (2000: 150)

In not knowing, the shape of justice to come is unpredictable, but that is not to say that present *injustice* is unrecognisable, or that our inter-est in it should be neutralised: 'On the contrary, it hyperbolically raises the stakes of exacting justice' in concrete cases 'with the most palpable effects' (Derrida 1990: 955). To fail to agitate for better law because the *avenir* of justice cannot be fully realised now – or ever – would be as eth-ically derelict as the assumption that justice is already installed in the juridico-political domain.

The question of ethics remains for Derrida the necessary corrective on the issue of law. Where the inherently transgressive singularity of the other demands a response and responsibility that cannot be codified in advance, one form that justice may take is that of an unconditional hospitality.[16] In his more recent work the two themes are inseparable, and Derrida's notion of the monstrous *arrivant* (1995: 307) has taken shape in the figure of the immigrant, the asylum seeker, or perhaps even the terrorist. Like those with transgressive bodies, all are paradig-matic of the stranger/outsider whose arrival is feared for the disruption it brings to socio-cultural and legal normativities, as well as to those of the subject. Yet to extend a welcome, as Derrida insists we must, does not devolve on the augmentation of rights, though they may be a pro-visional step, nor on assimilation to the values of the sociality. What matters instead is an openness to difference which responds '*without norm*, without presently presentable normativity or normality, without anything that would finally be an object of knowledge, belonging to an order of being or value' (Derrida 1995: 362). This is hospitality that owes nothing to the comfort of homogeneity or stability, and that is prepared to expose the standards of both the individual self and the sociality to the risk of the unknown and unforeseeable. It comes into play precisely where the conventional juridical and moral discourse reaches its limit. As Derrida notes:

> The law of absolute hospitality commands a break with hospitality by right, with law or justice as rights . . . . it is as strangely heterogeneous to it as justice is heterogeneous to the law to which it is yet so close, from which in truth it is indissociable.
>
> (2000: 25)

As such the principle of limitless welcome resists and transgresses the law that seeks to establish the *validity* of claims and in consequence always to set restrictions on when and to whom hospitality is owed. Derrida's alternative is a modality in which otherness is preserved, and the other interpellated neither to legal nor social subjectivity. It is the step that breaks with the Foucauldian cycle of power/knowledge with its endless recuperation of resistant difference to the ever-changing web of normativity. Instead, absolute hospitality is given before the other is identified. In short, undecidability persists – both in the form of justice or hospitality and in the irreducibility of the other – as the promise of ethics beyond the law.

By definition, the vision Derrida offers is unrealisable within the constraints of liberal jurisprudence or any other codified law, but his purpose nonetheless is not to dismantle, but to open up present-day structures by subjecting them to a critique that demands of them constant re-evaluation and renewal. As Brown and Halley note, critique is a way out of the impasse created when law supports a necessary liberalisation or extension of rights, and yet at the same time further enmeshes a particular group within the processes of normalisation (2002: 29). It is a timely reminder that in agitating for and celebrating certain legislative improvements, disabled people living in western societies will be all too familiar with strategies that although appearing to offer recognition of difference, nevertheless bring it under a rubric of control and regulation. Any simple appeal to legal rights around, for example, the sexual needs of disabled people – needs that may be casually thwarted by a normative model of sexuality that penalises the expression of more pertinent alternative modes of behaviour – cannot remain unproblematised. Current discussion within critical disability studies is struggling with the awareness that to simply legitimise access to sex workers, or the presence of a third-party 'assistant' in homosexual relations, or to take up the activist demand for sexual citizenship for disabled people would be to bring a yet greater part of disability under the scrutiny of governmentality. Moreover, the coils of heteronormativity are endlessly extendable, and pose uncomfortable questions to any radical sexual politics that wishes to acknowledge transgressive difference without domesticating it.

The specific question is whether juridical recognition can co-exist with a positive celebration of the otherness of the extraordinary body. There is no easy answer to such conundrums, but perhaps, as with the practice of psychoanalysis, we need first to speak what is repressed even in its irresolution, and then trace more productive paths from the legal unconscious to legal practice. Unlike the Foucauldian model, which can be read as accepting the inevitability of normalisation, the Derridean approach holds out a horizon of aspiration that paradoxically thinks beyond the achievable. It reaches towards that impossible justice for, that absolute hospitality to, the difference that maintains its singularity and strangeness. The coming of the monstrous *arrivant* – the one who is 'rebellious to rule and foreign to symmetry, heterogeneous and heterotropic' (Derrida 1990: 959) – will always transgress the law. The ethical task is to extend a welcome.

6
Queer Pleasures

As a fairly recent arena of academic growth, critical disability studies is committed to a questioning of the doxa of disability activism and scholarship in the interests of taking full advantage of the theoretical developments that have invigorated other areas of enquiry such as feminism, queer theory, postcolonial studies, and even critical legal theory. One part of my own project is to explore the possibilities of a Deleuzian approach, although there is as yet limited overlap between the material concerns of Deleuzian scholars and theorists of disability. What *is* by now well-established and growing, however, is a turn by many of the latter to the possibilities and insights of queer theory as an effective methodology for opening up a better understanding both of the relation between bodies, and of the constitution of corporeality in general.[1] The binary of disabled and non-disabled undoubtedly lingers within that approach, but it is increasingly destabilised by the intimation that all forms of embodiment are subject to reconstruction, extension, and transformation, regardless of the conventionally identified vectors of change and decay. What is more interesting, however, is the scope for a further productive move. The strong take up of queer theory within disability studies will lead, I suggest, to a reappraisal of the significance of Deleuzian notions like 'desiring machine', 'assemblage', and 'body without organs', all terms that have the potential to radically disrupt the devaluation of the disabled body. I do not mean to imply that the two perspectives are coincident, but that there are enough commonalities in their contestation of modernist conventions to encourage mutual fertilisation. By turning to minoritarian thinking/practices as recommended by Deleuze and Guattari, which seem highly appropriate to the problematic of disability, I want to explore not the shutting down or governance of sexuality, but the potentialities of effectively queering the terms of reference.

In this chapter, then, I shall continue with the project of questioning and contesting what it means to be a disabled embodied subject, and more specifically a sexual subject, by thinking very specifically about the material parameters that mark out which bodies are to matter. As previous chapters have developed at some length, the widespread anxiety in western societies that inflects the field of sexuality in general is at its most insistent and damaging when embodiment itself is perceived as anomalous. Whether the body in question has been intentionally transformed as in transsexual surgery or enhanced by body-building drug regimes, or has suffered severe trauma such as amputation or spinal injury, then the attributions of sexual desire and practice are most likely to invoke discomfort and confusion. And when it comes to disability, that disturbance may be heightened to become disavowal. The category of congenital or early onset disability is surely paradigmatic in that its exclusion from the very notion of sexual subjectivity is so underproblematised that it is taken almost as a natural fact. It is not necessary to re-essentialise sexuality, however, in order to contest the exclusionary violence of such a view. One option – which I shall go on to distort in a more productive Deleuzian manner – is to initially take the path forged by Michel Foucault and Judith Butler. Both make clear that what is at stake lies in the performitivity of sexuality, not as a potentially pleasurable bonus enjoyed by a pre-established subject, but as an uncertain process that infuses all aspects of the materiality of living in the world. For Foucault that uncertainty signals regimes of discipline and control which attempt to seek out and fix the 'truth' of sexuality, which far from being natural is actually an *effect* of hegemonic power/knowledge. While Foucault may appear to position the sexual self as primarily – though not entirely – reactive, for Butler, sexuality, though constrained by both socio-political and psychic dimensions, is nonetheless a core element of a more fluid self-becoming. In either case, sex is not something that bodies engage in, or in which subjects seek an identity: it is what constitutes us as embodied selves. And as such, it is not simply that sexuality may disturb the boundaries of the body or of identity, but that it is the locus in which corporeal identity both emerges – and is maintained through performance and management – and is simultaneously destabilised.

The implications of that insight are profound for those whose embodiment is non-normative, for whom sexuality is both devalued and denied, for it suggests that to silence or strip sexuality of significance is to damage the very possibility of human becoming. This is a matter of high ethical importance which devolves on the necessarily ambiguous relation between sameness and difference that cannot be resolved by a liberal humanist appeal to equality, not least insofar as that concept is fatally destabilised, as I have already suggested, by its implicit citation of a binary system of values that is defined by its dependence on, and hostility to, the non-normative. The issue, at heart, concerns the meanings and representations through which an embodied sexuality is constructed as a positive property of the normative subject, yet viewed as deviant, degraded, or simply not acknowledged at all, in the non-normative subject. Although my ultimate aim is to demonstrate the efficacy of a Deleuzian analysis in pursuit of an affirmative – indeed flourishing – account of disability and sexuality in which the latter is never simply a matter of agency, thwarted or otherwise, we cannot yet quite forget Foucault. As with most major postconventional theorists, Foucault largely overlooks the significance of disability,[2] but it is necessary, nonetheless, to revisit his work on uncovering the mechanisms in play in the construction and maintenance of the socio-cultural order. Despite some substantial signals of where 'bodies and pleasures' might subvert normative stability, Foucault is clearest in setting out the impressive array of disciplinary practices that are aimed at the singular body in all its aspects, but above all in its sexual pleasures (Foucault 1979a, 1980). As he shows, far from originating in an instinctual, biological ground, sexuality is always constructed through a highly complex process that is beyond individual agency:

> Sexuality must not be thought of as a kind of natural given.... It is the name that can be given to a historical construct: not a furtive reality that is difficult to grasp, but a great surface network in which the stimulation of bodies, the intensification of pleasures, the incitement to discourse, the formation of special knowledges, the strengthening of controls and resistances, are linked to one another, in accordance with a few major strategies of knowledge and power.
>
> (1979a: 105–6)

And, indeed, within Foucault's schema, bodies themselves are equally constructed, and thus open to endless transformation, rather than given entities. Nonetheless his thinking of how those bodies materialise has some curious omissions.

What Foucault notoriously fails to address – as Butler's powerful setting out and take up of the notion of performitivity (1990) makes all

too apparent – is the sense in which corporeality as sexed might be differentially constituted along the designated lines of male and female
morphology. But that is only the most obvious omission. What, we
are impelled to ask, of the enactment of other significant, and indeed
intersectional, differences, not least that which constitutes the binary
between ablebodied and disabled? It is not my suggestion that that particular division can ever be as clearly articulated as the one separating
male from female – although both conventional distinctions call for a
deconstructive analysis – but that there are similar urgent reasons to
interrogate the initial occlusion that covers over difference. In short,
and with due regard to the dangers of universalism, should we not conclude that the phenomenology of disability – with its potential absences,
displacements and prosthetic additions to the body – generates its own
specific sets of sexual possibilities that may both limit and extend the
performitivity of self-identity? If the normative standard against which
the acceptability of sexual practice is judged is male-dominant, heterosexual intercourse between two adults ideally acting without overt
external intervention, then in addition to a extensive range of familiar
refusals mobilised by preference, or at least some form of subjective decision, there are also certain morphological constraints that quite simply
preclude normative compliance. To have more or fewer limbs than the
norm, to be unable to hear, or see in the same way as the majority,
to have a prostheticised body, or to be conjoined,[3] are all conditions
that necessarily disrupt expectations of the 'proper' conduct of sexuality. It is not that any one of us – however we are embodied – can entirely
fulfil normative demands, and yet some forms of non-compliance evoke
not simply disapproval, but feelings of revulsion, albeit ones that are
threaded through with a certain fascination. The question of why the
morphology or comportment of some bodies should generate highly
negative reactions is just as significant with regard to the things of which
the body is capable and incapable.

At least part of the reason, which I have developed in earlier chapters,
emerges if the everyday socio-cultural understanding of what constitutes sexual normativity is underpinned by a more theoretical analysis.
To recap, what confer value in the modernist western conception of
the sexual subject are those familiar categories that establish autonomy,
that comprise notions of self-determination, separation and distinction,
and which demonstrate corporeal wholeness. And those are precisely
the qualities in which the universalised disabled body is deemed to be
lacking. Equally, it is those same categories that will be contested by a
Deleuzian approach. The problem for the modernist mindset, however,

is that in contradistinction to homosexuality, for example, which may offend against specific social mores around sexuality by failing primarily to perform appropriate models of masculinity or femininity, disability touches on a far more entrenched understanding of what it is to be a subject at all. Given that connotations of dependency and vulnerability – regardless whether they are operative or not – are understood to be antithetical to the attribution of full subjectivity in general, then the anxieties provoked by those qualities are all the more acute when their embodiment appears in a context that is already beset by all manner of putative threats to the autonomous subject. What I mean is that most sexuality is inherently about intercorporeality, about a potential merging of bodies, wills, and intentions, about a transmission of matter, and about an intrinsic vulnerability in which the embodied subject is not only open to the other in an abstract way, but is likely to be in a physical contact that is neither wholly predictable nor decidable. That the subject is never settled or simply present as a sovereign self, but intricately interwoven with the other in a dynamic process of self-becoming is, as I have shown, the basis of the phenomenological model of embodiment in a more general sense. But it is in the sexual relation, above all, that Merleau-Ponty's notion of the reversibility of touch (1968), with its implicit confusion of the boundaries between one body and another, and its potential for contamination, takes on a concrete materiality.[4] It is precisely because of the inherent risk of losing self-control and self-definition that the domain of sexuality is so highly disciplined and regulated, so saturated with performative constraints. And where the body of the other is already uncertain and resistant to the demands of normative comportment and expression – as it is paradigmatically in disability – then touch figures a moment of real threat, a troubling of the subject's illusion of purity and self-sufficiency.

The implication is not that the corporeality of disabled people is uniquely unstable, vulnerable, or interdependent, but rather that the nexus signals overtly what is more easily repressed in those whose embodiment satisfies the normative standards of western modernity. When Henri-Jacques Stiker refers to disability as 'the tear in our being' (1999: 10), he invokes a corporeal mode that – in the context of sexuality in particular – reveals the incompletion and lack of cohesion of the normatively embodied subject. And where the disabled body resists recuperation to the project of selfsameness, it figures not simply another, not like me, but becomes deeply disruptive of the very parameters that constitute selfhood. Its fate is to be refused any recognition in terms of sexual subjectivity, a refusal that typically takes the form of a silencing that intends a denial, and yet reveals, as Foucault (1979a) makes clear, precisely the complicity that it seeks to cover over. The very act of invalidation is both necessary and disruptive of the normative order. For Foucault, silence is an element of discursive power, but as I have suggested, it also indicates a psychic dimension to performativity that he leaves aside. Despite their explanatory power, then, it seems to me that neither the wider phenomenological approach nor the model of exterior governmentality is adequate to the theorisation of sexuality. Although Foucault convincingly charts the operations of a transformatory power over and through the body, even as one that is interiorised by each individual, he fails to take on the psychic significances of irreducible differences in embodiment. Judith Butler, standing as she does between a certain notion of phenomenology and the governmentality associated with Foucault, draws on the strengths of both in her own conception of performativity (1990, 1993), and given her attention to gender as a matter of discursive power, the step to disabled bodies is relatively straightforward. Butler is more explicitly insistent than either Merleau-Ponty or Foucault that any identity is performatively constituted only through doing, but given her parallel development of a sophisticated understanding of the unconscious processes at work in sexuality (1993), she has too a strong sense of other dimensions at work. In Butler's sense, the embodied disabled subject is not a pre-existing agent, but one that comes into being – and is materialised and sedimented – as a result of a series of acts and expressions.

My concern is that if body image – and especially internalised body image – is never simply a material reality but a complex and fluid mix of corporeal, psychic, and social components, then there is need for a more nuanced understanding, not simply of the *operation* of normative constructions of sexuality, but of the reasons for their emergence. In my attempts to theorise the question of disability and sexuality around such a problematic, I initially moved towards an analytic derived from psychoanalysis, particularly as deployed postconventionally. Despite its efficacy in uncovering the roots of the normative anxiety that grips that troubling conjunction, however, the paradigm seems to provide no way of unsettling a cultural imaginary that appears closed to a more positive model of corporeally anomalous sexual relationality. In other words, psychoanalysis critiques but does not fully queer the parameters of what is to count and what is to be occluded. Like women's sexuality, or more specific categories like lesbian desire, that have suffered a certain erasure in which the unsaid indicates an unthinkable anxiety, the conjunction of disability and sexuality is referred back to an explanatory model that implicitly privileges active phallic desire and the illusory quest for the restoration of an originary corporeal unity. I am not claiming that either the psychic or performative operations of gender and disability are directly comparable, but that both pose the question of whether any model based on the normative performance of male-dominant forms of genital sexuality has the capacity to encompass its excluded others. Having explored the seductive lure of the psychoanalytic approach and its inherent shortcomings in an earlier chapter, I will simply recall that trajectory here and move on to open up an alternative that retains a sense of psychic underpinnings, but owes more to Deleuze and Guattari than to Freud or Lacan.

As I understand it, the psychoanalytic model, which offers an explanation of the mechanisms by which the emerging subject moves from infantile to adult sexuality and is recognised as a sexual being, gives no real consideration to what difference morphological diversity would make. Aside from the supposedly inescapable biological sex of male or female organised around the materiality of the penis, other differences play a minimal part in the relevant theory. For both Freud and Lacan the acquisition and stabilisation of self-image is dependent on a certain corporeal introjection, not directly of the infant's own bodily boundaries and sensations, but of an ideal body image – for it is always in a sense an imaginary anatomy – representing both a mapping of the body's outer surface and an incorporation of the mirrored image of the other's body. In place of the maternal–infant dyad, the infant experiences a split which mobilises an endlessly substituted desire for that irrecoverable originary but undifferentiated wholeness. But if as Lacan implies, the putative unity of the self relies on the reflective unity of the specular other – indeed on jubilantly casting aside the infant's actual 'motor incapacity and nursling dependency' (1977b: 2) – then subsequently would not that new found sense of self be radically shaken by any mark of dis-unity in the external image? The disabled body, then, could be read as both insufficient as an object of desire, and an unwelcome intimation of the *corps morcelé* that the emergent subject must disavow or abject. It is not that the disabled infant would fail to negotiate the mirror stage – for in the psychic register all self-identity is based on mis-recognition. Rather, in its apparent lack of wholeness, the infant becomes other, its self-positioning as a subject of desire – like that of women – denied recognition. In such an account, the potential of difference to queer the terms of reference is effectively closed down. To escape the Lacanian impasse, I have undertaken to look elsewhere, and turn to a DeleuzoGuattarian alternative.

In decisively rejecting the Freudian/Lacanian model of desire as representative of, and mobilised by, lack, and by an implicit and impossible promise of completion and unity in a return to the mother, Deleuze and Guattari (1984, 1987) rewrite desire as productive, excessive to the embodied self, and unfixed. In their understanding, the psychoanalytic teleology of desire has no place. Rather than being goal-driven and singular, sexuality becomes, then, a network of flows, energies, and capacities that are always open to transformation, and so cannot be determined in advance. Where for Lacan, the *corps morcelé* of early infancy – and as I have argued the persistence of that body as a figure of disability – is seen as that which must be covered over in order to bring into being the unified self who will become a sexed and gendered subject in the Symbolic; Deleuze and Guattari celebrate precisely corporeal *dis-organisation*. The fragmented body is reconceived as the body-without-organs, the body in a process of corporeal becoming, that mobilises desire as a fluid indeterminacy that has no fixed aim or object, and which could always be otherwise. As Felix Guattari explains:

> For Gilles Deleuze and me desire is everything that exists before the opposition between subject and object, before representation and production. It's everything whereby the world and affects constitute us outside ourselves, in spite of ourselves. It's everything that overflows from us. That's why we define it as flow.
>
> (Guattari 1996: 46)

Instead of figuring the conventional ideal of autonomous action, separation and distinction, Deleuzian embodiment persists *only* through the capacity to make connections, both organic and inorganic, and to enter into new assemblages – which in turn are disassembled. Clearly the meaning of the body-without-organs is not intended as a denial of corporeality as such, but is rather a way of rewriting it that avoids the Lacanian narrative of a move from fragmentation to – at very least the illusion of – a temporally and spatially stable unity that grounds the subject. It is, then, the normative organisation of the body that is at stake here, an organism and organisation that closes down and fixes its possibilities rather than operating as 'a body populated by multiplicities' (Deleuze and Guattari 1987: 30). What Deleuze and Guattari want to promote is not a return to the staging of the pre-subjectival infant body, but a deconstruction, a queering, of *all* bodies that entails both 'taking apart egos and their presuppositions' and 'liberating the prepersonal singularities they enclose and repress' (1984: 362).

To think specifically of the disabled body in this context is not to single it out in its difference, still less to position it as inadequate. Rather it is a material site of possibility where de-formations, 'missing' parts, and prostheses are enablers of new channels of desiring production that are unconstrained by predetermined – or at least normative – organisation. Although the risk of stalling around an assumption of lack is always present, as it is with any body, the anomalous nature of disability holds out the promise of an immanent desire that embraces the strange and opens up to new linkages and provisional incorporations. And these connectors may make no distinctions between the human and animal, between organic and inorganic, or between substantive and semiotic flows. It is no longer possible to distinguish between an originary body and a prosthesis.[5] As Katherine Ott (2002) points out, the term 'prosthesis' has acquired rich abstract meaning in both psychoanalysis and cultural studies as a metaphor signalling some kind of mediation between an artificial device and the supposedly natural body, but it has also a complex material history mapping the literal interface between flesh and machine. I use the term in both senses, but want to stress the way in which disability may materialise some of the issues that underlie my concern with the performativity of the sexual self.[6] On the one hand, prosthetic devices are intended to replace or enhance normative function and appearance, figuring, in other words a Foucauldian sense of the technological disciplining and regulation of the body, but on the other, their use may be radically subverted.[7] The intercorporeality – or rather the concorporeality – of the organic and inorganic, the assembly and disassembly of surprising connections, the capacity to innovate, and the productive troubling of intentionality are all experienced by disabled people, particularly insofar as they are prepared to explore the uncharted potential of prostheses. As with other minoritarian thought and practices, like the feminine, the breaking through of the expected limits and constraints of the resources to hand can both intensify the decomposition of binaries – body/machine; active/passive; biology/technology; interior/exterior – and multiply non-repressive forms of passionate vitality. As Deleuze and Guattari note: '(d)esire constantly couples continuous flows and partial objects that are by nature fragmentary and fragmented. Desire causes the current to flow' (1984: 5).

In contradistinction to the Lacanian infant who after all greets its mirror stage *escape* from disorganisation with – as Lacan puts it – 'jubilation', this freeing up of desire, both in its object and its aim, may remind us of the polymorphous perversity of the Freudian infant, who finds undifferentiated sexual pleasure not only in every aspect of its own body, but in a variety of external objects.[8] As Freud points out, such perversity in the trajectory of desire persists in adulthood even in such everyday practices as kissing (insofar as it has no genital aim), but for the most part, it must be abandoned – repressed that is – not for the sake of psychic health, but in the interests of socio-political organisation. Nonetheless, despite the potentially productive tension that is set up by Freud's recognition that the price of such repression is neurosis (1962: 104), his reluctant turn away from polymorphous perversity shuts down precisely the queer reading of desire that Deleuze and Guattari are to reopen. It is sometimes tempting to think of Freud as the first – albeit thwarted – queer theorist, but Liz Grosz, in her own turn to a Deleuzian analytic, offers a less charitable view of polymorphous perversity. She warns against 'adopting the psychoanalytic position, which takes erotogenic zones as nostalgic reminiscences of a preoedipal, infantile bodily organization', or 'seeing the multiplicity of libidinal sites in terms of regression' (1995: 199). And it is precisely in the refusal to see alternative sexual pleasures as regressive that Foucault prefigures the queering of desire that is associated with Deleuze. Foucault's interest is both in what bodies can do – in how they are productive, rather than in how they respond to unconscious impulses – and in the ways in which the erotic can be redistributed to non-genital sites. In regard to S/M practice, for example, he is adamant that, '(t)hese practices are insisting that we can produce pleasure with very odd things, very strange parts of our bodies, in very unusual situations, and so on' (Halperin 1996: 320 quoting Foucault). Against the general societal condemnation of the perverse, disabled people who wish to assert an active sexuality might well find resonances in Foucault's words. I am not claiming that many would necessarily identify with the celebration of fetishism as such, but that the experience of a dis-unified or prosthetic body demands a degree of innovation and inventiveness that most of us rarely experience. For Deleuze and Guattari, that sexual creativity is surely at the heart of their anti-Oedipal project.

The conventional psychoanalytic approach that supports the normative postEnlightenment paradigm of a closed and invulnerable subject whose sexuality is organised around the presence or absence of the phallus, and whose sexual aim is to replace lack with plenitude, is supplanted, then, by a model whose potential positivity is unconstrained. In place of prohibition, repression and disavowal, Deleuzian desire is expansive, fluid, and connective, grounding sexuality itself as highly plastic and as no longer reliant on the terms of any binary opposition such as those of male/female, active/passive, or human/animal. And because the emphasis shifts from the integrity of the whole organism to focus instead on the material and momentary event of the coming together of disparate parts, bodies need no longer be thought as either whole or broken, ablebodied or disabled, but simply in a process of becoming through the unmapped circulation of desire. At the same time, desire itself takes on a wider meaning that liberates it not simply from the bounds of genital sexuality *per se*, but more generally from the restricted parameters of what is usually defined as *sexual* relationality, whether that is accepting of, or challenging to, the conventions. Skin on skin in the bedroom is no more privileged than the sensation of fine sand running through my toes, or the sweet taste of a juicy peach on my tongue. In an essay that is explicitly concerned to rethink lesbian desire, but which might equally open up the arena to the erotics of disability, Liz Grosz takes her cue from Deleuze and Guattari. She writes:

> there is not, as psychoanalysis suggests, a predesignated erotogenic zone, a site always ready and able to function as erotic: rather, the coming together of two surfaces produces a trading that imbues eros or libido to both of them, making bits of bodies, its parts or particular surfaces throb, intensify, for their own sake and not for the benefit of the entity or organism as a whole. They come to have a life of their own, functioning according to their own rhythms, intensities, pulsations, movements. Their value is always provisional and temporary, ephemeral and fleeting: they may fire the organism, infiltrate other zones and surfaces with their intensity but are unsustainable.
>
> (1995: 182)

Above all, what mobilises or stalls the rhizomatic proliferations of desire is the extent to which the connective nodules escape organised patterns of operation.

Desire is not an element of any singular subject; it is not pregiven; it is neither possessed nor controlled; it represents nothing; and nor does it flow directly from one individual to another. Instead it comes into being through what Deleuze and Guattari (1984) call 'desiring machines', assemblages that cannot be said to exist outside of their linkages and interconnections, and which may encompass both the animate and the inanimate, the organic and the inorganic. As Guattari puts it, we speak of machines, of 'desiring machines', in order to indicate that there is as yet no question here of 'structure', that is, of any subjective position, objective redundancy, or coordinates of reference. Machines arrange and connect flows. They do not recognize distinctions between persons, organs, material flows, and semiotic flows.

(1996: 46)

A desiring machine expresses, then, no necessary cohesion, continuity or unity, and nor do its part-objects seek a return to an originary wholeness, or find completion in an absent other.[9] What mobilise desire are not the endless substitutes for psychic loss, but the surface energies and intensities that move in and out of multiple conjunctions that belie categorical distinctions and hierarchical organisation. For Deleuze and Guattari, such conjunctions always engage the entire social and environmental field, centring not on the capacities of a unique individual, but on the scope and range of nomadic flows of energy – lines of flight – so that embodiment itself extends beyond the merely human. It is not that there is no distinction to be made between one corporeal element and the next, or indeed between the human and animal, or human and machine, but rather that becoming entails an inherent transgression of boundaries that turns the pleasures – sexual or otherwise – of the embodied person away from dominant notions of human subjectivity. As Tamsin Lorraine puts it: 'The self, rather than having a perspective upon and apart from the world of temporal becoming, is part of a process of dynamic differentiation' (2000: 185). This is not to deny that the interaction of bodies in time and space continues to produce subject effects, but it is only when those effects begin to coalesce and settle that the familiar sovereign individual of the postEnlightenment could be said to appear. The performative repetition of particular patterns and modes of organisation serve to construct an illusion of stability and permanence which is, nevertheless, undermined not only by what Butler (1993) sees as the inherent slow-motion slippage of all reiterative processes, but by the unruliness of the leaky bodies whose fluidity, energies, and contingencies are engaged in mutual transformations. These are bodies that come together – and break apart – in multifarious ways, always frustrating the anticipated outcome of performativity in consistent sexual identities. And where the stress is on the multiple possibilities of connection rather than on the putative dangers of contiguity and the risk of touch, then anomalous bodies are no longer a source of anxiety, but hold out the promise of productive new becomings.[10]

The stage is set, then, for a potential reclamation of disability and desire that is a very long way from the medium of an Oedipalised sexuality centred on the familial drama of 'mommy, daddy and me'. Like the female body, the corporeality of disability has widely figured in the western imaginary as disordered and uncontrollable, both seductive and repulsive, as threatening contamination of those who come too close, linked to disease, and so lacking in boundaries as to overwhelm normative subjectivity (Shildrick 2002). The link with sexuality is either disavowed or seen as overdetermined and abased, a matter of dangerous encounter that cannot but trouble the stability and self-presence of the unwary subject. That none of this would be articulated as such in the lives of disabled people is of little consequence. What matters is the power of the cultural imaginary to effectively exclude – in representative terms at least – a whole category of people from an important element in the socially normative process of self-identity. In contrast, what is offered by Deleuze and Guattari opens up a positive model of productive desire, the take up of which is limited neither to those who already fulfil certain corporeal criteria, nor to the sedimentation of a characteristically modernist form of autonomous subjectivity. It is a move from the fixity of being to the inventiveness of becoming. As Deleuze and Guattari explain:

> Starting from the forms one has, the subject one is, the organs one has, or the functions one fulfils, becoming is to extract particles between which one establishes the relations of movement and rest, speed and slowness, that are closest to what one is becoming and through which one becomes. This is the sense in which becoming is the process of desire.
>
> (1987: 300–1)

In place of the limits that the ideal of independence imposes on desire, the emphasis is on connectivity, and linkage. It is not that disabled people are unique in relying on a profound interconnectivity, but that where for the normative majority such a need may be covered over in the interests of self-sovereignty, it has come to figure a deficiency that ostensibly devalues those unable to make such choices. The disabled woman who relies on an assistant or carer to help her prepare for a sexual encounter – be it in terms of dressing appropriately, negotiating toilet facilities, or requiring direct physical support to achieve a comfortable sexual position – is not different in kind from other women, but only engaged more overtly in just those networks that Deleuze and Guattari might characterise as desiring production. Similarly a reliance on prosthetic devices – the linkages between human, animal, and machine – would figure not as limitations but as transformative possibilities of becoming other along multiple lines of flight.

How might this work out for disabled people who precisely because of their reliance on prosthetic devices would usually be characterised as dependent and unable to experience fully pleasure and desire? I want to look briefly at work by Barbara Gibson which sets out the practical day-to-day functioning of some severely disabled adults and theorises the observations using a Deleuzian frame.[11] By listening closely to the young men in her research study who all required long-term use of ventilators, Gibson comes to realise that the conventional therapeutic stress on achieving some form of 'independence' bears little relation to what the disabled men actually experience. She turns instead to Deleuze and Guattari's rejection of the fixed being of the subject that is defined by the *telos* of autonomy, and to their radical take up of the notion of active *becoming* that breaks through the bounded limits of the singular self. Gibson is never less than grounded, however, never in danger of losing sight of the actual young people she works with, and she makes clear that her subjects are 'both confined to individual bodies and simultaneously connected, overlapping with other bodies, nature and machines' (2006: 189). In illustration of how her new approach can make sense of lives as they are imperfectly lived, Gibson offers some examples of what she calls 'transgressive connectivity' that enable a re-imagining of 'disability' – the distancing quote marks are mine; the very word seems to be inadequate. One example – of a disabled man who uses a wheelchair, breathes with the aid of a ventilator, is nourished via a gastronomy tube and speaks through a voice synthesiser – might seem the epitome of conventional dependency, but Gibson sees it very differently. In his multiple connections, and the exchanges of energy that facilitate his capacities, the man is freed of the burden of individual identity to participate in a wider becoming:

> He is a fluid body, not a subject, but a conglomeration of energies. He has replaceable parts.... His organs are here, there, and everywhere.... He is an excitation, a point of contact, a relay on a power grid, a plot point on the plane of consistency.
>
> (2006: 191–2)

While the full account here is specific to a particular embodied person, I understand the Deleuzian-styled mechanisms that Gibson describes to connote both an individual moment of becoming through connection and a modality of existence for all of us. When Gibson notes that we all can, and will, move in and out of disability, she is not, I take it, simply making reference to the idea of the temporarily able body that is always at risk from illness and accident, but to a rather more radical acknowledgement of human vulnerability that figures an existential fluidity. But far from that signalling anxiety and restriction, vulnerability itself opens up the possibility of desiring production through an intense connectivity.

A further example, again taken from Gibson, serves to illustrate the point, this time in the specific context of disability and sexuality. I have referred above to the way in which a disabled person might experience her sexual encounters as moments of connection, not only to her partner, but to any assistant who might be involved in the scenario as a facilitator. What is different about Gibson's approach here is that she focuses not on the disabled employer, but on the '3rd party', the one who is supposed somehow to be both present and absent at the scene of sexual intimacy. After quoting an assistant's reported awareness of her own contemporaneous sexual concerns, Gibson comments that nonetheless, '(t)he attendant is expected to be a detached "tool" for facilitating their coupling – a means to an end' (2006: 192). As I discussed in Chapter 4, facilitated sex is in any case a site of much anxiety, a discomfort which must surely be heightened by the contradictory demands put on the assistant, who in the contractual relationship of independent living may be even more silenced than the employer. The tension is clearly caught in a poem entitled 'To My Other Bodies' by the disabled writer Connie Panzarino (1996), who speaks of her 'strange' and intertwined relationship with her attendant who is at once an extension of her own body, and yet who must affect invisibility when the poet is 'alone' with her lover.[12] Panzarino has few illusions as to the blurring of boundaries between the two women, but that is still to overlook the embodied participation, the slippage and fluidity of identity, and the exchange of energies experienced by the paid employee herself. The problem, as Gibson astutely notes, is that, despite knowing her role, any *assistant* 'experiences a leaking of her identity, a mingling of her own sexuality with theirs' (2006: 192). Gibson's understanding of the multiple encounter goes much further, I think, than the phenomenological intertwining of self and others that I have previously outlined and takes up a more fully Deleuzian account, not just of mutual construction, but of dynamic becomings that rely on profound connections: '[Even] without a sexual *act*, the two (the three) intermingle on multiple strata, becoming one–(an)other' (*ibid*). What this intends is that the bodily transgressions associated with disability – which are in fact open to anyone – are a powerful step towards opening up the unlimited potential of living through, becoming as a result of, the dynamic and always unfinished processes of assemblage. Desiring machines speak to all manner of couplings, organic and inorganic alike, that push the experiences of embodiment to a different plane that is as hospitable to disabled people as it is to any others. Indeed, insofar as disabled people may have less invested in the trope of sovereign subjectivity, or are better able to let go of such illusions, it may be more so.

We should caution, nonetheless, against taking an overly romanticised view of disability in which desire is always able to operate as an unchallenged positivity. In advancing their anti-Oedipal thesis, for example, Deleuze and Guattari (1984) rhetorically evoke – or appropriate, as Prendergast (2008) would have it – the schizophrenic to figure the dis-organised, multi-faceted, fluid, and unpredictable nature of all that stands against the formalisation of the conventionally embodied subject. But what of those who in everyday life bear the diagnosis of schizophrenia: is their condition *experienced* as a positivity? Prendergast does not deny the efficacy of the postmodernist challenge to the normativities of social and personal order and identity, and recognises that disability studies has much to gain; what concerns her instead is that Deleuze and Guattari – like many other postconventional scholars – are making of the schizophrenic figure something exceptional that bears little connection to lived reality. Reviewing the increasingly public presence of schizophrenic people as part of the socio-political emergence of disability more generally, she remarks: 'they claim the right to *unexceptional* instability, which is not something postmodern theory has readily granted them' (2008: 57, my emphasis). Clearly the power to speak out as schizophrenic, rather than being spoken of, is an important step to recognition in the public sphere, but I am not sure that posing the problematic in terms of rights can ever escape the limits that socio-political acceptance brings with it. If to be unexceptional means to be indistinguishable from everyone else, then does not that yet again cover over the irreducible difference that all forms of individual embodiment encompass? Once again the complex layering of what constitutes disability, and therefore how it might be better responded to, seems to be lost. Although Deleuze and Guattari may indeed have lost touch with the 'reality' of what they purport to describe, there are others who speak from just such a devalued position and yet present their personal difference as a matter of energy and productivity. I am thinking particularly

of the recent work of James Overboe (2007a,b), who lives with cerebral palsy, and who utilises an explicitly DeleuzoGuattarian mode to
celebrate the strange capacities of his body. He speaks unequivocably of
their work as having pragmatic sociological application to real lives, and
writes of his own condition: 'As a person who experiences cerebral palsy,
the affirmation of (spasmodic) desires could result in a flight from the
restrictive humanism that confines my expression of life to the facile
categorization of disability' (2007b: 28).[13] For Overboe, it is not a matter
of simply transgressing the norms of the social order, but of entering
into the absolute immanence of *exposure*, a state that operates as though
the norms did not exist.

There are clearly some constraints, some morphological differences
and discontinuities, that continue to impede the flow of energies, particularly if that flow has been mapped in advance. But the model I
propose here is not about such modernist notions as unrestricted choice,
or about a freedom that opens up all and every possibility. Like Butler's
original exposition of performativity (1990), which was widely misunderstood to offer unbounded access to self-stylisation, the notion of
desiring production must always be contextualised. The rewriting of performativity as intensely connective, and the slippage of reiteration as a
more radical discontinuity, highlights precisely a lack of control that
may exacerbate the frustration of intentionality for some people with
disabilities. But rather than offering a route to sexual identity, the model
proposes something rather different: a break with the putative emergence of a coherent sexual subject from the practices of embodiment,
and a turn to the libidinal intensities which play not across unified and
integrated bodies, but at points of connection between disparate surfaces or entities that may or may not be organic. The desire produced
in and over the dis-organised body owes little to genital sexuality or the
goal of self-completion in sexual satisfaction. As Liz Grosz notes:

> desire is an actualisation, a series of practices, bringing things
> together or separating them, making machines, making reality. Desire
> does not take for itself a particular object whose attainment it
> requires; rather it aims at nothing above its own proliferation or
> self-expansion.
>
> (1994a: 165)

and elsewhere she characterises the appeal and power of such desire as
'its capacity to shake up, rearrange, reorganize the body's forms and sensations, to make the subject and body as such dissolve into something

else' (1995: 204–5). It is not that Deleuze and Guattari allow no place for subject effects – 'you have to keep small rations of subjectivity … to enable you to respond to the dominant reality' (1987: 160) – but that those effects are unsustainable in fixed form, beyond the temporary or provisional. The molar politics of identity and subjectivity are never entirely dismissed, but are constantly confronted and displaced by the molecular politics of flows and intensities. What matters to Deleuze is the transformative potential of the process of becoming. In being excluded from full sexual subjectivity, then, those who are disabled have lost nothing of permanent value.

What this all indicates is that were the western privileging of autonomous individuality and integrated identity less rigid, the performativity of (sexual) subjectivity could be vitally transformed. Despite its commitment to the productive instabilities of discursive construction, the notion of performativity remains focused on a form of individual agency that might be more radically queered by taking account of the emergence of the self precisely through an erotics of connection. As Guattari notes: 'desire is everything that exists *before* the opposition between subject and object, *before* representation and production' (1996: 46), and Deleuze himself goes further in his deconstruction of the relationship between a willing agent and desire: 'Far from pre-supposing a subject, desire cannot be attained except at the point where someone is deprived of their power to say "I"' (Deleuze and Parnet 1987: 89). As more and more theorists are beginning to acknowledge, the corporeality of disability is not that of an other fixed in a binary relation to the normatively embodied self, but is already queer in its contestation of the very separation of self and other.[14] The so-easily silenced whisper of a kinship that would be denied – for it unsettles the foundations of western subjectivity – is growing into a roar that marks a new understanding of embodiment which owes much to Deleuze. Having now entered 'the next century' of which Foucault (1977b) claimed Deleuze as the philosopher, I should like to offer the equally bold speculation that the Deleuzian project will be realised at least in part through the medium of rethinking disability.[15] Once again, it is not that disability is a unique case, but only that its forms of embodiment, and its embrace of prosthetic enhancement, seem to overdetermine the fragility and instability of corporeality in general. The postmodernist acknowledgement that all bodies – normative and non-normative alike – are in a constant process of construction and transformation, brought about not least through interactions in the spatio-temporal dimensions of the social world, means that all are potentially hybrid, nomadic, machinic assemblages. In the specific differences of its capacities – particularly with regard to its libidinal investments – the disabled body exposes, moreover, the queerness of all sexuality.

That disability should be perceived as dangerous, and that its erotic capacities should be disavowed, speaks to the threat that it is able to unsettle the normative constraints that attempt to limit adult sexuality to a highly regulated set of impulses that cover over the rhizomatic operation of desiring machines. For the most part, the libidinal possibilities of surprising, unpredictable, non-respectable, even dangerous conjunctions, which are in principle open to all of us, are kept in check by the rigid and repetitive structures of a normative sexuality that cannot easily countenance unauthorised variation or experimentation. To limit the erotic to the law of desire as it operates within the hegemony of the Symbolic is to assent to a system that can give no adequate account of corporeal difference nor of an alternate sexual imaginary, yet to be realised. It is to close down on fluidity, on connection, and on intercorporeality, and to impose prohibition or denial on those who are assigned to positions of social marginality. For people outside the mainstream, then, those who are transgendered, HIV positive, or disabled, the choice may be between either an apparent asexuality that comforts normative expectations in its very powerlessness to mount a challenge or an expression of desire that will be necessarily exploratory and transformative. It is not of course that all disabled people are sexual radicals or have any urgent wish to liberate their desire from the constraints of normative thought and practice; as for all members of regulatory societies, it is impossible to stand outside the networks of disciplinary power/knowledge by any simple act of will. The performance of sexuality is never an unimpeded choice. Nonetheless, as comparative outsiders, many such minoritarian figures are already engaged in a queer performativity that takes off from the innovative and intimate connections that are often a necessary part of life with a disability.[16] As I have discussed, such expansive operations both encompass the many forms of personal assistance that are available in the West, and which inevitably entail an embodied relationality that goes beyond normative encounters between putatively autonomous selves, and accommodate the enormous range of prosthetic devices that already may be incorporated into the experiential field of a person with a disability. For others, it is not so much a matter of describing present practice but of thinking otherwise about the promise of connectivity, and about what would follow from attending not to the being of a subject, but to the becoming and doing that constitutes a provisional and contingent subjectivity.

In all cases, as Goodley and Roets put it: 'We need to rename the sexed and impaired subject as a multiple, open-ended, embodied and interconnected entity in a network of simultaneous power relations' (2008: 246).

The substantive specificity of disabled bodies nonetheless poses something of a conundrum for those willing to deploy a Deleuzian rather than more generalised queer approach to the problematic of bodies that matter. Where queer theory explicitly intervenes in the parameters of social exclusion – and to an extent must always reiterate binary thinking in order to contest it – it readily lends itself to the critique of mainstream socio-cultural values and regulatory norms that disabled people engage with. The idea of assimilation is thoroughly rejected, and as Foucault recognised: 'It is the prospect that gays [read: *disabled people*] will create as yet unforeseen kinds of relationships that many people cannot tolerate' (1997: 153). As such the lure of identity politics lingers on, bolstered by queer theory's oppositional take on normativity that has broadened out to include differences in race, ethnicity, and potentially embodiment, as well as sex and sexuality. The embrace of multiple and diverse minorities in queer theory undoubtedly casts the dominant standards in an ambiguous light that troubles the centre, but it also appears to gesture towards the reappearance of a stable subject. In contrast, when Deleuze and Guattari refer to minoritarian practices, they are not ontologising any given category as an identity; instead 'all becoming is a becoming-minoritarian' (1987: 291). Moreover, their notions of becoming-woman, becoming-animal, or becoming-minoritarian are not simply conceptually unattached to the groupings named and open to all, but refer to processes that operate only through the assemblages temporarily brought about by radically disparate machinic connections. Where, then, does this leave the substantive minorities, like people who are disabled? From a DeleuzoGuattarian perspective, they too must enter into the process of becoming, a process that entails both contesting the relations of power that structure every fixed subject position and leaving behind any existing modes of identification. It is risky and uncomfortable, but necessary insofar as the disabled body, however well-adapted or accommodated, is – like the figure of 'woman' to whom Deleuze and Guattari consistently refer – always constituted by the repressive organisation of modernist principles.

So long as all fixed and unified identities rely on the performative exclusion of an abject domain of the unthinkable (Butler 1993), then certain bodies will never matter. If on the other hand, our mutual and irreducible connectivity were recognised as quite simply a condition of becoming, and as the ground for the positivity and productive play of desire, then the notions of independent agency and self-containment that mark the normative subject might lose their exclusionary power. In place of the liberal demand for rights, choice, and self-determination that presently shape the dominant discourse of disability activism, a more open and productive model that celebrates the qualities of those already living at the margins might be proposed. It is their very disorganisation, and their necessarily overt contiguities with an array of others, that the better enables such figures to breach the boundaries and explore what lies beyond the normative limits. As Sara Ahmed notes: 'The unreachability of some things can be effective; it can even put other worlds within reach' (2006: 153). To go further, an openended and ambiguous, yet more positive Deleuzian mode of becoming has significant implications, not only for the hitherto disavowed conjunction of disability and sexuality but for everyone. The point is to move away from the notion that desire represents or substitutes for an originary loss that codes all bodies in the same way; instead it maps the multiplicity of becoming. And once the plasticity of the erotic is acknowledged rather than repressed, then the circulation of desire and the partial satisfactions of pleasure would be a matter of differential exploration and experimentation, rather than the site of silence and shame. There are of course dangers – not all lines of flight will soar – but the possibilities of reconstruction and transformation, in sexuality as elsewhere, speak to the hope of personal and social flourishing. That term is deeply unfashionable within postconventional discourse, but it is precisely what I mean.

7
Global Corporealities*

In this chapter, the queering of corporeality in the mode of disability will be continued primarily through using the work of Deleuze and Guattari, which will be brought together with some crucial elements of feminist theory, and placed in critical conjunction with the context of globalisation. In previous chapters, a number of illuminating pairings have already emerged from those dimensions, but in focusing on the context of the traditional West – or in global terms, the socio-economic North – little has been said of how the problematic of disability, subjectivity, and sexuality plays out in the wider world. In part I have exercised a cautious reticence against claiming too much, but given our increasing awareness of the influence of global forces on all aspects of life, it makes sense to at least offer some more extensive speculations. My starting suggestion is that if the more usual approach to the nexus of disability – usually traced through personal and local socio-political experience – is opened up to take account of global interconnectivity then a very different form of analysis is needed. In the face of a limited but consistent literature that takes a traditionally materialist and pragmatic outlook on the intersection of disability and global politics,[1] I shall instead turn once more to the insights offered by Deleuze and Guattari, and by postconventional feminism, in order to expose the conventional approach to a theorisation that more fully accounts for heterogeneity and differential embodiment.

An important aspect of my approach has been throughout to stress the intersectionality of disability discourse with other fields that speak to embodied subjectivity. If disability and globalisation have rarely been theorised together, the greater surprise is that feminist scholarship, which has been at the forefront of rethinking corporeality in all its guises, has had relatively little to say about forms of embodiment that are categorised as disabilities. And despite the range and strength of feminist activism over the last few decades, both locally and globally, there appears to be a similar reluctance to confront the issues at the heart of this book. New and often exciting work is emerging all the time, but it still feels necessary to pose the question: why should feminism take disability seriously? Given feminism's central concerns, why is it not self-evident that disability will always be on the agenda? As is well-documented, contemporary feminism links back to a history in which notions of equity are paramount, and since the second wave, it has produced some highly sophisticated work on reclaiming the political, socio-cultural, and indeed philosophical importance of embodiment. Moreover, after a fairly shaky start in the 1970s and 1980s, it has come – albeit 'kicking and screaming' as Donna Haraway put it with reference to white Anglo-American feminism (1991: 155) – to recognise and finally to celebrate the significance of multiple irreducible differences that take in race, ethnicity, and sexuality, to name just a few. In addition, the domain of disability itself displays a very significant gender dimension, the acknowledgement of which might alone seem to guarantee that the problematic constitutes a legitimate concern for feminism. Recent WHO (2004) estimates, for example, indicate that over 600 million people worldwide experience disability,[2] but it appears consistently that the impact of disability on women – and especially those living in the socio-political South – is overdetermined, regardless of whether the proximate cause is poverty, disease and ill-health (including HIV-AIDS), labour-related accidents, warfare and militarism, or congenital factors (Boylan 1991; Hershey 2000; World Bank website 2004). Unfortunately, neither such pertinent empirical features, nor the ongoing theoretical context of feminism, has ensured that the issue of disability has been widely taken up in women's studies, in either the North or the South.

One major hindrance is that the problematic is seen almost exclusively as the domain of those with personal experience of disability, of interest to feminist theory only in passing. Certainly it remains very unusual to find relevant work by non-disabled feminists. In a highly nuanced consideration of what she sees as a clear lack of awareness of the intersectionality of disability and feminism, Rosemarie Garland-Thomson makes the point that the difficulty lies on both sides. As she remarks: 'many disability scholars do not know either feminist theory or the institutional history of Women's Studies.... Conversely, feminist theories all too often do not recognize disability in their litanies of identities that inflect the category of women' (2002: 1–2). Whatever the intentions of the author, Carol Thomas' book, *Female Forms* (1999), is a case in point. Although it explicitly discusses feminist ideas in the context of the social model of disability, it is published not as feminist theory but in a series called 'Disability, Human Rights and Society'. Moreover, even where there is acknowledgement of a legitimate field of enquiry – in disability studies or disability politics itself – feminism has struggled to make its concerns heard. As Anita Ghai, an Indian disabled feminist, points out, 'men and male concerns have dominated' (2003: 51). Nonetheless, despite that gloomy assessment, there are some clear signs of a new and growing awareness that disability is not just another add-on concern. In other words, some recent analyses have positioned disability as a central rather than marginal component in better understanding the way in which socio-cultural, psychological, and political vectors operate in the contemporary world. It is not only that disability could arguably be named as paradigmatic of inequality in general, that it is not just one difference among many, but a difference that changes everything. It is my own strong contention that disability affects every one of us whatever our personal corporeal form, because our mode of embodiment is one – if not the major – organising principle by which we make sense of the world.

As I indicated in the preceding chapter, a combination of queer and Deleuzian theory has begun to emerge in critical disability studies as the site of the transformative resources most able to open up the terrain.[3] There is already growing scholarship that utilises the two with regard to disability, but the point is that they are applicable far beyond their own putative origins in sexual critique. Like feminism, the application of queer theory has no obvious disciplinary limits, and lends itself not only to a rethinking of discrete elements, but to an interrogation of the links between them. In particular, and in deep contradistinction to LGBT studies with which it is still sometimes confused, it has deeply problematised the meaning of identity and identity politics. Indeed, it is the commitment to flexibility and hybridisation on which queer subjectivities are premised that stages a disruption of profound significance to the convention of (identity-based) disability rights. Any deployment of a fixed – albeit positively valued – identity as a challenge to socio-cultural intolerance, or as the basis for rights claims, runs the risk of affirming the very difference that generates oppression in the first place.[4] Moreover, it rejects the potential of the new connections that are opened up by queering mo(ve)ments. In the face of that closure, several theorists have recognised that the coming together of queer theory and disability studies can shift the problematic to a more productive phase that embraces, rather than denies, the inherent instabilities and vulnerabilities of the embodied self that disability can so readily exemplify.[5] In place of identity, the emphasis rests more appropriately on performativity, on self-becoming not as an intentionality within the control of a singular agent, but as a process that is neither free nor fully determined. Where Butler's (1990, 1993) seminal account has some-times been read as downplaying interpersonal context in favour of the reiterative operations of custom and law, both components are intrin-sic to the queering of identity that the notion of performativity signals. The antecedents of its interconnective processes are already evident in the phenomenology of Merleau-Ponty, but it is the further move to a Deleuzian analysis that more fully enlists the energies and intensi-ties of multiple transformatory conjunctions.[6] Thought together, queer theory, disability, and Deleuze mobilise a productive positivity that over-comes normative binaries, breaks with stable identity, and celebrates the 'erotics of connection' (Shildrick 2004b).

How then is the movement of globalisation – which in the early twenty-first century has become an apparently unstoppable force – relevant to the question of disability, and more importantly suscepti-ble to an analysis that queers both its epistemological *and* substantive bases? One major starting point – and there are of course many others such as those looking at the material effects of capitalism for example – is that the context of globalisation radically alters any understanding of bodies, and of the place of those hitherto situated on the margins of what is valued. Globalisation can mean many different things, but what I have in mind here is its significance as a constellation of material and imaginary, spatial and symbolic processes that constantly intertwine one with the other. Over and above the more familiar politico-economic meanings, this somewhat more expansive connotation is important because it signals that where *material* globalisation is characteristically and justifiably taken to be partial, exploitative, and elitist, those other less substantive elements may have more positive aspects. In short, it cannot be assumed that the critique of globalisation proceeds only in a negative mode. My own purpose, then, is less to highlight the social, political, and economic inequities involved, than to trace out the phenomenological dimensions – the experiential nature of living-in-the-world-with-others – of the global coming together of bodies. At the very simplest level, this has powerful implications: if – as I have con-sistently argued – we are all irreducibly situated in a *network* of corporeal relationships, then the whole notion of a fixed centre and margins must be problematised. What this suggests is that to pass judgement on glob-alisation as though its value or disvalue could be simply read off its concrete effects on discrete categories of the rich and poor, ablebodied and disabled, consumers and producers is an overly facile approach. A far more nuanced and open approach that is sensitive to symbolic as well as material aspects is required.

There is no question that a commitment to contest any *solidifications* of power in the operation of globalisation remains necessary, but the point is that such outcomes are by no means definitive or exclusive. An equally compelling parallel need should not be ignored, and that is to engage with the expanding and intrinsically unstable frontiers of what Michael Hardt and Antonio Negri term a 'decentred and deterritorialising apparatus of rule', through whose networks the global flows of 'hybrid identities, flexible hierarchies and plural exchanges' can be mapped and explored for their positive potential (2000: xii). Indeed, as they suggest in their quasi-Deleuzian analysis, power is not simply dispersed but, by the same mechanism, constantly challenged by a reverse movement: 'Resistances are no longer marginal but active in the center of a society that opens up in networks; the individual points are singularized in a thousand plateaus' (2000: 25). Along with Hardt and Negri, many other commentators have noted that one major marker of what is understood as *symbolic globalisation* is that in the era of postmodernity we experience a quasi-condensation of space and time to the extent that the world appears to shrink.[7] While this could be taken to imply a certain universalisation that covers over difference, it might alternatively signal, as Hardt and Negri suggest, a destabilisation of existing hierarchies and the collapse of any one normative standard. Rather than assuming, then, that the already existent and historical marginalisation of many forms of corporeal anomaly would be simply superseded in the processes of globalisation by a yet greater occlusion of difference, it may be just as likely that the projected condensation would afford new opportunities to rethink the relations between different forms of embodiment, as well as encouraging new ways of performing identity in a world of others. The link back to the phenomenology of intercorporeality, and to Merleau-Ponty's wonderfully evocative phrase 'the flesh of the world', is strong.

When Janet Price and I first decided on the title 'That Shrinking Feeling' for a presentation at the 2004 Women's Studies Association conference on feminism and globalisation, we were thinking in terms of how the speed of movement and change in a contemporary world dominated by global capital is instrumental in *undermining* the familiar boundaries of separation and distinction by which those who are valued are held apart from those who are not. There is indeed a sense in which the shrinking horizons are not about newly imposed limits and constraints, but just the opposite: that we are witnessing an opening up of apparently limitless possibilities of both connection to, and familiarity with, difference. The effects remain complex, signalling both homogenisation and differentiation – even fragmentation. And despite the association with the emergence of the worldwide market of twenty-first-century capitalism, there is inherent within the connective structures of globalisation the very configurations that enable a productive critique of its effects. At its most optimistic, and contrary to the view that the speed of globalisation brings about alienation and a loss of specificity and authenticity (Doel 1999), it is possible to think – as even some writers for whom the language of queer theory or postmodernism means nothing – in terms of a positive exposure to new cultures, new bodies, and new people.[8] But there is of course another connotation to that shrinking feeling which we have explored in previous work: namely the impetus of those who are normatively embodied to shrink away from encounter with those who are not. It is readily apparent in the disinclination of those who are putatively ablebodied to literally touch people who are anomalously embodied, or to be touched – physically and metaphorically – by them (Price and Shildrick 2001). That shrinking away is a distancing move that insists on distinction and separation, and it is evident not only in the personal encounter, but takes the form of a wider refusal to acknowledge the significance, and both local and universal implications, of differential embodiment. What that shrinking away indicates is, of course, a denial of the phenomenological constitution of all embodiment whereby none of us is given to the world as whole, autonomous, and self-present.

Given such tensions, what are the possibilities of taking up the theme of connectivity that is implied both by the metaphor of shrinking time and distance, and, as I go on to outline, by DeleuzoGuattarian notions of becoming, of assemblage and of the rhizome? In *A Thousand Plateaus*, the specific image of shrinking holds a privileged place in the project of becoming-imperceptible (1987: 279), a move which signals not so much the disappearance of individual agency as its liberatory dissolution into multifarious microintensities. But how can we avoid pushing supposedly marginal figures – whose agency is already compromised – beyond having *any* place in a reworked network of spacio-temporal global exchanges at both the material and symbolic level? If the potential benefits of globalisation are to be realised, then they must not come at the price of an unawareness of the many dangers to those already without power or recognition. One immediately evident shortcoming is that it is only the relatively well-off – those who are busy and short of time, but in control – who can live in a perpetual present, whilst the poor, who also often busy and short of time, must always struggle with a lack of control of both their past and their future, as well as with the immanence of place. In other words, nomadic shifting (which Deleuze and Guattari in another register have theorised as the privileged mode of becoming), the sense of being a space/time traveller, appears to be only fully open to those with resources. Having money facilitates such a move, or access to the internet, or owning a mobile phone to text messages, commands, and demands from 'here' to geographically, and conceptually, distant locations. Clearly such possibilities are not free of power relations, but they do expose the intricate and multiple pathways that thread between any two points or persons, disrupting the structural hierarchy of centre and margins. In the Deleuzian world, the usual points of reference are put aside, and the linearity of time and the regular expansion of space are superseded by a virtual mobility characterised by speed, intensity, and energy operating in what Doel calls a 'scrumpled universe of general relativity' (1999: 14).

A second major difficulty is that for those able to experience that possibility of being everywhere and nowhere, corporeality – the very condition of a grounded connectivity – seems to slip away. Many disabled people, for example, take advantage of multiple chatrooms and dating websites, precisely because cyberspace – in appearing detached both from physical constraints and from pre-existing visual description – encourages the restaging of personal details and specific forms of embodiment. Although the protocols of online dating state that users should be truthful in how they present themselves, in actuality, in a world freed from the demands of the local and the material, users can put on any character they choose. As Juniper Wiley writes, 'Newly generated personas – faceless, voiceless, bodiless – displace history with a timeless present and multiple selves easily co-exist with the flick of a finger. Fantasy is freed' (1999: 135). That such online performativity has become so acceptable and commonplace indicates not so much that technologies may be radically transformative, but rather that specific corporeality is always unstable and open to queering. But does this mean that in the dephysicalised time–space compression of globalisation, where spatial dimensions and temporal gaps are collapsed into singular moments – as Bauman puts it, 'the shrinking of space abolishes the flow of time' (1998: 88) – bodies no longer matter? Have they become, in effect, everywhere and nowhere all at the same time? On the contrary, time–space continues to exert a differential friction on bodies. For all that web-masters and online currency dealers are quasi-decorporealised, they also rely on the all too fleshy bodies of global service providers – transport operatives, farmers, healthcare personnel – to ensure that their own bodily needs are met. It is never simply a matter of acceleration, speed and compression, but of an inevitable collision, and complex interplay of worlds over such mundane matters as the delivery of an ergonomic office chair, the availability of a favourite lunchtime coffee blend, or the need to get new spectacles. 'Everywhere it is a matter of speed and slowness, of territorialization and deterritorialization, of foreshortening and elongation' (Doel 1999: 188). Whilst it is clear we can play with and disrupt corporeal expectations, it is less evident that we can do away with bodies altogether. It is not just that a transparent and weightless present/ce is not available to all, it may in any case be only an illusion. However advanced, technological augmentation may obscure the ground of materiality, but it relies finally on corporeal input, however that is staged. Embodiment finds new forms, but it does not disappear.

On a more material level, the phenomenological approach developed by Merleau-Ponty has already made clear how bodies and selves come into being and are continually reconstructed through concrete encounters and contact with those around us, those whom we touch or with whom we are in touch in the course of our everyday lives. According to the phenomenological understanding of embodiment, which I outlined in Chapter 1, we are all in a reversible relationship – both touching and touched, perceiving and perceived – with a range of others from whom we are never wholly detached. The feminist philosopher and bioethicist Ros Diprose has both taken up the insights of Merleau-Ponty and extended them to take account of global corporeality, a notion that might seem beyond the latter's remit and yet is inherent in his own deployment of the term 'the flesh of the world'. Assuredly Merleau-Ponty was not thinking of globalisation, but he nonetheless moves towards an appreciation of both the all-encompassing geographical dimensions of intercorporeality, and – as I indicated in my earlier reference to Rod Michalko and his dog – the possibilities of non-human encounters. Diprose does not take up that latter element – which will lead us via phenomenology to Deleuze – but concentrates on the move from the immediacy of embodied experience to the wider canvas of our embodied response and responsibility to those others who are beyond the here and now encounter. If I exist as a social being only insofar as my body is in mutual interaction with others, then my corporeal connectivity seems to speak directly to moments of *local* encounter and relationality that ground a particular form of ethics. What Diprose (2002) names as 'intercorporeal generosity' intends not some quasi-contractual exchange between sovereign individuals, but an openness of the embodied self to embodied others. In the immediate field of cross-cutting affectivity and perception, my interconnection is clear, but it does not resolve the question of what limits might restrict the range of bodies with which my gestures of intercorporeal generosity can engage. Even in many daily activities, we have moved beyond the point where the encounter is face-to-face – I am as likely to email a colleague as walk into her office – but, although it is more diffuse, the connection between us, and the call to an ethical relation, remains strong. What is more complicated is the issue of how my own embodiment meshes and interacts with bodies that are *not* part of a familiar environment. What are the dimensions and implications of my intertwining with those far away, those whom I will never know, and perhaps can scarcely imagine? As I sit writing in my study at home in the United Kingdom, what affect do I have on, or connection with, the bodies of countless unknown others dispersed throughout the world, or they on me?

The recognition that there are irreducible connections already in place, however much they are daily pushed aside, is one consequence of the analysis of globalisation. And even as they circulate through a network of social, political, and cultural mediators, those connections impact on specific corporeality in such a way that, rather than reducing the issue to a question of individuals, queers the very nature of individuality. Consider, for example, a simple tracing of the demands of the western consumer through a chain that links bodies to bodies. In the previous chapter, I referred to Barbara Gibson's account of a disabled man using a variety of prosthetic devices such as a ventilator, voice synthesiser and wheelchair, all of which rely on a flow of electricity that links him indirectly to the hydroelectric dam. Gibson goes on: 'energy is exchanged with other bodies that maintain the dam, manufacture the machines, and service the body' (2006: 191). The global dimensions are immediately apparent. And Price has reflected in a similar way on 'the complexity of corporeal interdependence that underlies my use of a car to carry both myself and my wheelchair around' (Shildrick and Price 2005–2006: ¶10). Her listing of those who might be involved in simply supplying the car – from French designers, through Filipino piece workers, Indian call centre workers to English dealers – involves a global collaboration of workers in both the economic North and South, rich and poor, unionised and casual labour alike. The network of connections could almost be extended indefinitely. For Price in particular, the specificity of her car must reflect the needs of her changing corporeality, and, as she recognises, the vehicle is never a simple utilitarian mode of transport, nor even a phenomenological extension of her body, but forms what Deleuze and Guattari would call a new assemblage with it. Like the example of the disabled man that Gibson develops, the accounts of such extensive connectivity throw into question the nature of the distinctions between self and other, and between independent and dependent. In each case, a disabled person gains what appears to be enhanced *self-dependency* precisely through dependency on others, who at the same time are at least partially dependent for their own well-being on his or her consumption of goods and services. More importantly yet, it should be apparent that although individual requirements may vary greatly, every one of us functions only through such assemblages. Reflecting on the complex hybidisation between human beings and technological devices, and using Haraway's language of the cyborg (1991), Gibson notes: 'The cyborg is connected to the cosmos, no more or less so than all bodies.... all contemporary persons are connected cyborgs' (2006: 192).

The usual reason to draw on such examples of consumption is to mark out the one-way flow and reach of western power and its negative impact on the bodies of others. Whilst not denying the ever-present potential of harm, the point in this chapter is quite different and requires of the reflective reader a willingness to set aside any call for immediate action, or corrective policy initiative, whatever form that might take. As already indicated, my concern is less with the globalisation of inequality – which many others are engaged in mapping – than with the no less important task of thinking through the phenomenological significance and implications of global intercorporeality. If the capacities of any one body alter in relation to any other body – a disabled woman's mobility in response to a Filipino woman's manual dexterity; her diet in response to the Westerner's reliance on a specialist car – then it can no longer be assumed that bodies exist in discrete spaces. Even as a member of an affluent society, such a disabled woman is nonetheless relatively disadvantaged and may not be able to exercise the full power of the ideal consumer, yet, like everyone else, she is necessarily situated in a web of embodied relationships in which the configuration of centre and margins is deeply problematised. There are two significant and interconnected aspects to this. First, the whole transaction represents the very Foucauldian notion of the dispersal of power,[9] where even those whose privilege is relative – having access to personal transport will always be both less than and more than some others – are never wholly marginal

nor central to the processes of consumption. At the same time, the mul-
tidirectional linkages also represent the irreducible bodily connections
through which the corporeal productivity of the one is intertwined with
that of others. The separate and distinct categories of self and other,
exploiter and exploited, non-disabled and disabled, bosses and workers
all become blurred, in turn making any fixed evaluation of the effects of
globalisation open to question. While any persistent imbalance remains
potentially dangerous and should be challenged, the point – even in
relation to the dynamic and often unpredictable circulation of power –
is not primarily to make value judgements. The task, more importantly,
is to trace the modes by which new configurations of time and space
are operationalised, to uncover the flows of energies that globalisation
enhances.

In the light of such concerns, much of my own past collaborative
work with Janet Price on disability (Shildrick and Price 1996; Price and
Shildrick 1998)[10] has taken off from the Foucauldian concept of normativities that both offers a genealogical analysis of the stark binaries
that organise postEnlightenment thought, and undoes those opposing
categories by revealing their discursive construction. In addressing the
global, however, the work of Deleuze and Guattari is more productive –
because as Deleuze explains, '[Foucault] . . . was establishing novel historical sequences, while we put more emphasis on geographical elements,
territoriality and movements of deterritorialization' (1990: 150). More
particularly, the global interweaving of corporeality that is axiomatic
of a potentially more positive approach to the issue of globalisation
finds strong resonances in the way in which Deleuze and Guattari work
through the notion of fluid connectivity and linkage. It is precisely those
rethinkings of the individual and the social that facilitates the moves
that encompass both the near and the far, those in direct contact and
those whom I will never encounter. The question that has been posed
and left hanging – 'Can there be any notion of corporeal generosity
over distance?' – may find an answer here. But before moving on to
consider some of the ethical dimensions of rethinking disability in a
connective context, it is useful to briefly recall the wider significance of
using a Deleuzian approach. Aside from some isolated references, disability theorists, as I have already noted, generally have little place for
Deleuze and Guattari, although it is notable that both Rosi Braidotti's
early work on Deleuze and later on monsters (1991, 1994) and Liz
Grosz's (1994a, 1995) on 'volatile' bodies indirectly indicate how the
approach might be incorporated. For my own part, I have no compunction in raiding the theoretical toolbox to assess whether the Deleuzian

analytic offers any insights that might be adapted to the end of further queering the problematic of disability and globalisation. Like Braidotti and Grosz, I want to explore new articulations of the body. My aim is to provide neither an unassailable mode of reasoning that is true to the originary authorial intention, nor the comfort of absolute clarity, but to provoke further exploration along new lines of thought.[11] In any case, Deleuze (1995) himself calls for an active, productive engagement with existing theory and concepts that may be developed or recreated in quite different fields. Above all, my use of Deleuze represents not an end in itself, but a challenging starting point for reconfiguring what is at stake.

As I have already set out in the previous chapter, Deleuze and Guattari (1984, 1987) decisively reject any understanding of socio-cultural organisation that depends on notions of lack or wholeness, or of sameness and difference; that relies on separation and distinction, on hierarchy, or on putatively necessary repressions and prohibitions, justified as the price of a well-ordered society. In contrast, their emphasis is on the positivity of desiring production, which arises in the flows, energies and intensities of nomadic wandering, in hybrid associations, in the acceptance of ambiguity, and above all in an ever-expansive connectivity in which not human beings as such, but human becomings, are but one element. What has been off-putting for many feminists and radical disability theorists alike – those few that is who are prepared to engage with such postmodernist insights – is that on a superficial reading Deleuze and Guattari appear to reject not only the materiality of the body, but also the specificity of particular forms of embodiment, the very thing that has driven feminism and disability studies. Certainly, their widely misunderstood figure of the body without organs appears to support such a contention, but their purpose is otherwise. For Deleuze and Guattari, the notion of BWO intends not a turning away from corporeality as such, but rather a deprivileging of organisation. The deadly point of organisation is a corporeal unity and integration that stands in the way of acknowledging the profound and complex linkage not only between diverse human beings, but between humans and animals, and humans and machines. Instead, the DeleuzoGuatterian term, the body without organs, figures a dis-organ-isation that will open up myriad unpredictable and temporary lines of connection and encounter. As Liz Grosz puts it:

> The point is that both a world and a body are opened up for redistribution, dis-organization, transformation; each is metamorphosed in

the encounter, both become something other, something incapable of being determined in advance, and perhaps even in retrospect, but which nonetheless have perceptibly shifted and realigned.

(1995: 200)

The highly irregular network of connections that the Deleuzian model proposes – explicitly figured as a 'horizontal' rhizomatic proliferation of linkages rather than the 'vertical' and therefore hierarchical image of arborescence (1987: 16–17) – offers too an intimation of inclusivity that is very different from Bauman's (1998) understanding of how globalisation might work. Like Deleuze and Guattari, Bauman is strongly critical of the structures of modernity, but he warns that in a newly mobile world of potentially instantaneous communication and access, the dephysicalisation of time and space is highly dangerous for those unable to take advantage of its promises. As he sees it, acts of separation and the establishment of hierarchies will become even more merciless, precisely because not everyone can be part of a mobile elite that crosses boundaries at will. In particular, those already located at the margins – people who are poor or dispossessed, asylum seekers, those who are disabled or somehow hybrid – are all at risk of being permanently kept in place. As Bauman remarks: 'Rather than homogenizing the human condition, the technological annulment of temporal/spatial distances tends to polarize it' (1998: 18), and again: 'The mark of the excluded in the era of time/space compression is immobility' (1998: 113). Such pessimism is by no means easy to dismiss, but for those attuned to the ethical significances[12] – indeed the necessity – of multiple manifestations of difference, the charge that technology *fails* to homogenise human life will scarcely be regretted. It is not so much that Bauman himself is indifferent to difference, but that his anxiety is misplaced. For a most postconventional critique, and above all for queer theory and contemporary feminist scholarship, the point is that to count any homogenisation of the human condition as positive would be as unlikely as wishing for the hierarchies of further polarisation. The issue, on the contrary, is to preserve and recognise difference without solidifying hierarchy. And what that entails is not some reification of balanced diversity, but a queering of difference in the form of provisional and asymmetric hybridities.

It is precisely with regard to these concerns that the turn to Deleuze and Guattari may signal another more positive meaning to high-tech global operations in the era of postmodernity, and one that, despite current suspicions or disinterest, might be particularly appealing to disabled people. In offering at least a partial counter to Bauman, I want to explore how the model proposed by Deleuze and Guattari could be applied to some of those groups of people whom Bauman believes are most at risk of experiencing the negative effects resulting from globalisation. Given that what is at the heart of the DeleuzoGuattarian project is a celebration precisely of corporeal dis-organisation, it is not difficult to flesh out that abstract notion in terms of the lives of disabled people. Rewritten as the body-without-organs in the Deleuzian sense, the disabled body can bypass separation and distinction, and demonstrate a persistence that both respects and exceeds its own specificity by making fluid and open connections in multiple directions. In a model in which corporeality is no longer to be thought in terms of given and integral entities, but only as engaged in ever dynamic and innovatory linkages, bodies are neither whole nor broken, disabled nor ablebodied, but simply in a process of becoming. And the point is that the process follows no set pattern, nor has any specified end. There are, then, no fixed hierarchies, nor predetermined limits on the nature or trajectory of the connections to be made. It is not that there is no distinction to be made between one corporeal element and the next, between one human body and another, or equally – for Deleuze and Guattari – between the human and animal, or human and machine, but rather that becoming is a process of ever-new and always provisional points of coming together. It creates what they call 'a zone of proximity ... a non-localizable relation sweeping up two distant or contiguous points, carrying one into the proximity of the other' (1987: 293). Above all, it entails an inherent transgression of boundaries that turns the capacities of the embodied person away from normative categories of inclusion and exclusion.

If we follow Donna Haraway's dictum that '(q)ueering specific normalised categories is not for the easy frisson of transgression, but for the hope of livable worlds' (1994: 60), then the ethical implications of such a process are always there to be excavated. But is it possible to lay out more explicitly what the Deleuzian notion of becoming might mean in terms of global relations. Once again Ros Diprose's approach suggests a way of bringing together the diverse concerns of this chapter with an ethical framework that goes way beyond the liberal humanist desire for a level of equality across all the differences that might distinguish my body 'here' from your body 'over there'. Though we *are* all caught up in reciprocal relations, whether we acknowledge them or not, the mere fact that we mutually impact each other's corporeality does not seriously challenge, still less eliminate, asymmetrical power. But that uncomfortable reality provides no obstacle to ethics, for as Diprose suggests:

> intercorporeal generosity maintains alterity and ambiguity in the possibilities it opens, it is not based on an ideal of mutual exchange between equals.... generosity is only possible if neither sameness nor unity is assumed as either the basis or the goal of an encounter with another.
>
> (2002: 90–1)

What this implies is a form of ethical responsibility – a corporeal generosity – that in the face of globalising tendencies both preserves and challenges the import of the interval – a space of not-knowingness – between the putative self and other. As a matter of ethics, it is incumbent on us to maintain that tension, to rethink the whole significance of nearness or distance. It is a thought that goes against any attempt to simply construct an overarching – yet finally problematic – sense of empathy as that which would seem to supersede for the better the conventional understanding of distance as a modality that lends itself to a lack of care, engagement, or trust. Such empathy always implies a degree of knowingness, a grasp of the other that suggests a certain violence. In constructing her model, Diprose takes little from Deleuze and Guattari or queer theory as such, and yet her phenomenological and ethical concerns imbricate with both. The reconfiguration of difference common to them all, and more specifically for Diprose, the rethinking of ethical responsibility towards global differences in the era of postmodernity, requires not the modification or extension of existing paradigms, but their deconstruction.

It would be a mistake to suppose, then, that the shrinking of distance implied by globalisation directly instantiates an ethical relation. Nonetheless, what the changes in our phenomenological relation to space/time compression, and the fluidity of the configuration of centre and margins, should do is to make us rethink the question of ethics. Elsewhere (Shildrick 2002), I have posited an ethics of encounter to supersede the fatal limitations of a morality based on the distinction between self and other, but it needs to be clear that the encounter may remain virtual and never be realised in concrete terms. As Liz Grosz puts it in one of her Deleuzian commentaries, '(e)thics is the sphere of judgements regarding the *possibilities* and actualities of connections, arrangements, lineages, machines' (1994b: 197, my emphasis). In the shift and flow of relationship, we need to recognise both the mutual constitution of self and other – whether that other is near or far, *and* that difference is never fully reducible, that some element of it will always remain apart and incommensurable. Ethical generosity, then, is less a matter of the knowingness of empathy than of acknowledging the unknown and unknowable dimensions of relationship. The point is that whilst the dynamics of globalisation encourage an appreciation that hitherto marginal figures are not so distant, that is not to reduce them to a commonality. What is called for is an ethical recognition and consideration of those very differences that are not easily reclaimed to a singular standard, but still maintain an interval and resist the all too seductive grasp of the move of empathy. Disability is just one of the discomforting differences that has to be thought, both in its ubiquity in every aspect of socio-political existence, and in its strangeness and capacity to elude the predictability of the known. Where on the one hand non-disabled and disabled people alike are co-implicated in a dis-organised flow between both themselves and other others, the non-normativity active in that intercorporeality can elicit also a break, unforeseen lines of flight, a moment of difference, within the interrelation between bodies.

Several of these issues become clearer in the context of a personal account offered by Janet Price, who speaks of experiencing a serious relapse of Multiple Sclerosis that resulted in a temporary paralysis that required the daily use of support workers. All were employees of an outsourced local agency which recruited people without much regard for prior experience or training, and which paid the minimum wage. Inevitably, a number of the workers were from economically disadvantaged contexts, and many had come to the United Kingdom as refugees or as students, who, having few choices, were prepared to take on what the well-off may see as menial employment. Price continues:

> In the course of the episode, my embodied self-identity was severely disrupted as I struggled to re-form myself in the new experience of both the changed structures of my body, and the unfamiliarity of the relationships with those around me. What becomes critical are the linkages with those carers whom the shifting territories of globalisation had brought into my living room, bedroom and bathroom. All sorts of differences were at stake, which our unanticipated modes of connectivity could not cover over. One of the care workers, for example, a Sri Lankan woman, once asked whilst she was showering me, 'Why don't you cut your pubic hair?'. Although we were undoubtedly intimately connected intercorporeally – her hands on my naked body – her words had a profound affect, making me feel anxious and

disturbed. What for me brought up questions of whether I would ever be able to be sexually active again, probably seemed to her – coming from a different cultural context – a straightforward query about the maintenance of bodily hygiene and propriety. For both of us, the moment caught the strangeness of coming together, the instant at which self-familiarity is unsettled and opened up to new modes of becoming otherwise.

(Shildrick and Price 2005–2006: ¶17)

For the most part such fleeting moments of strange encounter may pass unacknowledged, the resistance to disturbance too great. But *potentiality* – in Deleuzian terms, the inherent power of assemblage – relies not on the comfortable fit, but on just such asymmetries:

Everything that happens and everything that appears is correlated with orders of difference.... the condition of that which appears is not space and time but the Unequal itself, disparateness as it is determined or comprised in difference of intensity, in intensity as difference.

(Deleuze 1994: 222–3)

On a wider level, the whole issue of intimate bodily care – which falls within what Hardt and Negri (2000) term 'affective labour'[13] – raises some highly pertinent questions: what is its relation to becoming(-with-others); how should it be addressed alongside the new techno-benefits available to disabled people that are opened up by, and figure, the speed and de-materialisation of the postmodern world? Just as Deleuze and Guattari claim, the linkages that matter here are not just those between human beings, but between all manner of animate and inanimate forms brought together in temporary assemblages. But that is not to dismiss the enduring place of human-to-human touch, albeit in networks of connection that stretch from near to far. While machines can type letters, turn the lights on and off, open doors, and cook food on a timer, what such techno-servers cannot do is help to bathe, to dress, and move disabled people in the most intimate of ways. The people – and they are usually women[14] – who provide such services are often part of a nomadic chain of support: the woman studying in western Europe, to enable her to earn enough to support her family, works part time for an agency providing personal care, but leaves behind her young children in the city back in her home country. They in turn must be cared for by an itinerant rural woman who can no longer support herself on the land and so travels daily to the city from the village, leaving behind her own children to be cared for by elderly grandparents (Ehrenreich and Hochschild 2003). Flows of personal support and of finance pass along this chain, which marks a familiar pattern of global labour. But the exchange of money is not the only thing at stake: the hundreds of thousands of women worldwide, who are drawn into the care economy, at both a legal and an illegal level, exchange countries, families, modes of living, and bonds of affection. The intercorporeal care and assistance that many disabled people experience in daily life has significance far beyond the confines of personal comfort. Those encounters stage a process of becoming that speaks to a rhizomatic proliferation of connections, situated in nodular global networks coalescing in temporary points of assemblage. Regardless of how any one of us may be socially located, such interactions – direct and indirect – profoundly reshape our subject being, and ultimately the dimensions of the worlds we inhabit.

The materialities of those interconnections function in other ways too. Many of us from more secure backgrounds, who travel to less economically privileged parts of the world – and indeed to less privileged parts of our own cities – wonder in good liberal ways how we should respond to those whom we encounter begging, many of whom are disabled in one way or another. The limits to such lives, their relative immobility, stand in stark contrast to the tourist style wanderings that may bring us face to face. In many societies and religious faiths, including western ones, there is a strong tradition of dispensing charity to the poor, although as Henri-Jacques Stiker points out, the idea of assistance mainly exists where there is no analysis of poverty as socially produced (1999: 127), or, in the contemporary world, globally produced. But that, in any case, bears little resemblance to the possibility of generosity which concerns me here. Is the disabled woman with her baby selling perfumed flowers to passing drivers at the traffic lights really so apart from those she appeals to? It is certain she too has a life and family behind her, of which most of us know little, but there are other more disturbing linkages. Her injuries may be those of an overstretched agricultural field-worker picking the peas for my dinner party, or of a child stepping unaware on an abandoned land-mine from some past or present war, waged in the interests of the West; her untreated congenital condition or AIDS may reflect the profits of multinational drug companies – those same companies that provide me with travel sickness medication; her arthritic hands the result of years of labour in a garment sweatshop that produces logo-laden leisure-wear. It is all too easy to think no further than the figure of oppressor and victim, but the turn to Deleuze opens up a different approach. Although he is never unaware that globalisation may be just another aspect of control society, he also notes: 'It's not a question of asking whether the old or new system is harsher or more bearable, because there's a conflict in each between the ways they free and enslave us' (1995: 178). The issue, then, is not just one of uncovering the existing networks of power that have constructed us both, but of recognising further, more positive, dimensions of our intertwined possibilities of becoming. The gesture of giving individual money – even where expected and welcome – may serve to solidify the fixed relation between giver and receiver, closing down corporeal generosity in Diprose's sense, *or* it may mark a new assemblage. In my present bodily engagement with the other – whether it be open and expansive, or defensive and anxious as I rapidly wind up my car window against the impoverished appeal – and in my acts of recognition, I can open out, or severely limit, the possibilities of an intercorporeal connection that continuously restructures identity – both the other woman's and my own.

The implication is that although it is all but impossible to escape complicity in the asymmetries of power, the move to change the stress from inequity to focus instead on interconnection mobilises new ethical possibilities. This is no easy matter to accomplish for there is much to overcome: in the *western* imaginary at least, disability is figured as dangerously out of order and beyond the subject's intentional control; as both an object of fascination and repulsion; and as always a potential challenge to categorical boundaries (Shildrick 2002). For the modernist mind with its stress on order and organisation, these are all highly negative modes, and, as such, the encounter between the normative and anomalous body is figured not so much in terms of a power relationship that could be realigned, but as intrinsically troubling and dangerous. Alongside such discursive violence, the outcome is material damage in that the entrenched power of the imaginary is sufficient, as I have shown, to cast a designated category of people – the bodies that do not matter – beyond the confines of normative socio-cultural engagement and socio-legal protection. For Deleuze and Guattari, however, the way forward lies not in the modernist discourse of political struggle, the worn out offspring of liberal humanism, but in the move to promote dis-organ-isation in all its aspects, and to offer a virtual model of 'desiring production' (1984). It will not be the agency of a self embodied in a complete and integrated organic unity that is the driving force of socio-political change, but the flows of energy that bring together part objects – both living material and machinic – to create surprising new assemblages. In place of the limits that the ideal of independence imposes, the emphasis is on connectivity, and linkage. Even were the move only one from the limitations of actuality to the potential of virtuality, it demands that disability be rethought.

The Deleuzian canon makes no specific reference to disability, but it scarcely needs to.[15] Despite the profound privileging of independent agency in the domain of western hegemony, the complex and never fixed interconnectivity that may overtly characterise a mode of life recognisable to disabled people encompasses, as Deleuze and Guattari understand it, every one of us. The fear of dependency is sufficient to effect a denial by those who see themselves as autonomous subjects mastering their own corporeality, but in the end the perceived form of embodiment is not the difference that determines who shall flourish. Given, moreover, that significant linkages operate variably between the organic and the inorganic, any recourse to prosthetic technologies must figure not as ameliorative responses to existing deficiencies but as a mode of enhancing the transformative potentialities of becoming other along multiple and unpredictable pathways. It is not a case of simply adding one element to another for the sake only of an additional functionality that leaves the component parts unchanged, but of effecting a new assemblage that is as unrecognisable and unexpected as it is energetic and productive.[16] For Deleuze, the generative power of life is rhizomatic, a horizontal proliferation of nodules and connective channels that eludes the more familiar pattern of growth and development through aborescent branching that preserves a central origin which remains essential to the whole. The rhizome 'is composed not of units but of dimensions, or rather directions in motion .... When a multiplicity of this kind changes dimension, it necessarily changes in nature as well, undergoes a metamorphosis' (Deleuze and Guattari 1987: 21). The prosthetic extension of the body always exceeds the boundaries of the embodied subject, making connections through other people, animals or machines whose own becomings mobilise yet other lines of flight. This is no challenging new dimension peculiar to the mode of disability, but an aspect of life familiar to all. It is difficult to draw a meaningful distinction between, for example, a disabled person's use of a voice synthesiser and the growing use of text messaging, between reliance on a assistance dog and riding a horse, between the amputee who uses a wheelchair cart and the executive in a speedboat, or between the wearer of corrective orthopaedic boots and an athlete's high-tech sprint shoes. All extend the bounds of possibility by making connections – by both organic and technological means. Moreover, the ubiquity of the prosthetic body is not limited to the era of postmodernity; the body has always engaged in connective transformations, as Haraway puts it, 'queering what counts as nature' (1992: 300).

Despite the affirmative nature of the alternative approach I am suggesting, there is still a need – as I warned in Chapter 6 – to guard against speaking of the disabled body as though its capacity to enter into productive assemblages were always assured. Corporeal differences may be experienced in a variety of ways – including pain and exhaustion – that obstruct the flow of energies, even as they also contribute to deterritorialisation. Certainly the demands and expectations of rehabilitative medicine, that still plays a major part in disability discourse, understand pain and exhaustion as signifying primarily personal limit(ation)s to be overcome. Moreover, the political and socio-cultural constraints, with which disabled people are all too familiar, will not simply disappear. Although the potentialities of the disabled body can be written into, or more radically – through prosthetic linkage – mobilise, webs of rhizomatic energies, they may also speak, then, to an experience of blockage or lack of self-determination. Indeed Deleuze and Guattari themselves acknowledge that even creatively dis-organised flows of energy can coalesce – precisely in those kinds of socio-political organisations that may serve to devalue and thwart disabled people. Yet the point here is that the normative status of autonomous action and personal invulnerability are simply the illusions of a rationalist modernity that the Deleuzian model seeks to supersede. What matters to Deleuze is the power to affirm life, both in its uncontested moments of joy and in its endurance, and that is precisely what Overboe catches when he writes of 'the vivaciousness of cerebral palsy as a life affirming force' (2007a: 221). It speaks not to the strategy of transcendence as a way of overcoming bodily disarray – and in any case Deleuze was uninterested in functional efficacy as such – but to what Rosi Braidotti has called 'sustainability' (2006: 29), the capacity to enhance one's potentials through pleasure and pain alike. Both speak to a heightened sensitivity. The task of the individual, then, is to embrace without *ressentiment* all that occurs, as Deleuze puts it, 'not to be unworthy of what happens to us' (1990: 149). The good life is one that overflows and transforms itself even in the face of adversity, always opening up new possibilities of becoming other than itself. And for Deleuze, like Braidotti, what a body can take far exceeds normative expectations.

There is, besides, a further consideration that may turn an initially negative assessment of bodily 'impairment' to positive account. There is a strong sense in which it is the perfectly ordered and healthy subject who is least able to enter into the multiple becomings that affirm life, while those identified as vulnerable or deviant are well-placed to contest the sovereignty and deadening sameness of what constitutes normativity (Braidotti 2006). In a similar way, Hardt and Negri claim that the transformation of corporeal relations rely on the body 'that is radically unprepared for normalization'. Although they concede that the evolution of such transformations remains as ambiguous as it is necessary, they insist – with credit to Guattari – that if globalisation (in its negative sense) is to be met with counter-globalisation, then the anomalous body is a crucial site of positive resistance:

> The will to be against really needs a body that is completely incapable of submitting to command. It needs a body that is incapable of adapting to family life, to factory discipline, to the regulations of a traditional sex life, and so forth. (If you find your body refusing these 'normal' modes of life, don't despair – realize your gift!)
> (Hardt and Negri 2000: 216)

It is just such gestures that have met with the criticism that Hardt and Negri have failed to appreciate the depth of material oppression associated with global capital,[17] but that is too flat a reading: on a symbolic level, their claim is well-founded, and the more striking for being counter-intuitive. What is put in question, both here and yet more clearly in Deleuze and Guattari's approach, is the privileging of independence, which in any case is not a cross-cultural given, but a specific outcome of a very western way of understanding the encounters between subjects. Instead, then, of holding out the unattainable ideal of self-control and independence, the model speaks to a reconfiguration of corporeal relations, and of time and space, that constantly produces the new and unexpected.

In taking up, and to some extent wilfully reshaping, a Deleuzian mode of thought, a radical transformation in the meaning of disability becomes possible. The task is one of shifting the focus to encompass a very different approach that starts with a radical questioning of organisation and structure *per se* and of the requisite closures that they perform. Indeed, it is in the very dis-organisation of those already living at the margins, and in their necessarily overt intercorporeality with an array of others, that it is possible to find new directions in a globalised world that is rejecting many of the values of western modernity. To be positioned at the margins of normative structures may be figured less as a failing than as an opportunity to breach outmoded bound-aries and explore what lies beyond. Such an analysis surely resonates strongly with the feminist deconstruction of masculine hegemony since the second wave of the 1970s, and as a gender studies scholar as much as a critical disability theorist, I see convincing reasons for develop-ing further intersections between the two perspectives. The mutual suspicion of postmodernist approaches is, I believe, unwarranted and reflects a lingering desire for an illusion of certainty and identity that no longer makes sense. In any case, as Rosemarie Garland-Thomson argues, feminism already 'tolerates internal conflict and contradiction' and simultaneously 'asks difficult questions, but accepts provisional answers' (2002: 25), and to a large degree, queer theory too – particularly where it embraces a Deleuzian perspective – has already shed much of the baggage of the past and taken up the challenge of thinking dif-ferently. It is time for disability theory to learn from the evolution of postmodernist feminist and queer theory. The unreserved commitment to ambiguity and contradiction, and to a project of disturbing existing paradigms without promoting stable or enduring alternatives, is espe-cially open to the task of developing a new, and one hopes ethically alert, responsiveness to the transgressive corporeality of disability – both in and out of its global context.

   Using the theoretical concerns of Deleuze and Guattari to reposition minoritarian interests begins to mobilise an effective account of glob-alisation that incorporates symbolic and imaginary aspects without losing touch with bodies. More particularly, it suggests that globalisa-tion not only offers opportunities for new locations, energies, capacities, and technologies, but opens up different forms of fluid identity that encompass – indeed see as paradigmatic – the lives of disabled people (and others) who are currently marginalised. This is emphatically not a programme that will guide change, but an exercise in thinking otherwise. Of course there are all too many empirical obstacles, but it is not necessary to see them as a limit on our endeavours. Although care is always needed to recognise the complexity of global differ-ences, the problematic of disability demands of us all a responsibility to follow through on the adventure of rhizomatic thinking. As the cen-trality of the western notion of the individual subject is disrupted by global forces, ethics itself is taken apart and reassembled to become, in Braidotti's words, 'a mode of actualizing sustainable forms of transfor-mation. This requires adequate assemblages or interaction: one has to pursue or actively create the kind of encounters that are likely to favour an increase in active becomings' (2006: 217). Both Deleuze and Braidotti are clear that such an approach is apposite for a highly technologised society in which our encounters – and our ethics of encounter – encompass not the other as another liberal humanist subject, but an unpredictable array of those dissimilar to ourselves. And as the world shrinks, it calls on us to exercise a corporeal generosity that acknowledges both our claims on, and our debt to, all the others – human, animal, or hybrid life forms – whose multiple ways of becoming intersect with our own.

Conclusion: Thinking Differently

> We have to learn to *think differently* – in order at last, perhaps
> very late on, to attain even more: *to feel differently*
>
> (Nietzsche 1982: 104)

The postconventional and often explicitly postmodernist theories that I
have been using as my mode of analysis throughout are proving essential tools to critical disability studies because they enable us to embark
on the crucial ethical step of thinking differently. Nietzsche might not
be the most obvious philosopher to associate with refiguring disability,
but his project of thinking, and consequently feeling, beyond normative
frameworks sets up the conditions of possibility for a radical socio-
cultural *and* socio-political revaluation of who and what is to count
in the era of postmodernity. As I have been arguing all along, critical disability studies takes its lead from a plethora of postconventional
theories – which in the main have little or nothing to say directly about
the specificity of the disabled condition – and creatively bends and rear-
ranges them to reveal new insights pertinent to the issues in hand. It
is simultaneously an act of *bricolage* and resistance that aims to break
through the limitations of the modernist paradigm to construct more
appropriate modes of engaging with the disabled body. Whilst acknowledging that the concerns of liberal humanism have contributed greatly
to the amelioration of the material oppressions directed against disability and cannot be lightly dismissed, a more critical approach takes the
further step beyond the straight-jacket of binary differences to explore
the fluidity of all forms of categorisation. To think from the field of
the other – people with disabilities – becomes, then, not an exercise in
standpoint theory that would create new hierarchies of privilege, but

a way of thinking otherwise that strives to exceed the very experience of boundaries. Nor is it a matter of seeking out a single approach that will better answer to the demands and hopes of disabled people, but of creating the possibility for questions, directions, and breakouts as yet unthought.

The contrast with the rhetorical strategies of conventional political activism and even much emancipatory discourse – give us our rights, now! – could not be greater. I am not suggesting that those struggles are either unproductive or outmoded, and should be abandoned, for there is still much to gain by challenging the normative political, juridical, and social structures of any society that practices categorical exclusion. Nonetheless, I am committed to the view that broadly postmodernist alternatives both more effectively analyse and deconstruct the structures that maintain those damaging normativities, and mobilise new and more creatively positive ways of thinking and feeling about difference. What many critics find intolerable is that the question of what comes next is deliberately left open, and whether disability studies takes on the work of Derrida or Deleuze, there are no easy answers. There is, in short, no programme or plan of action, no right or wrong way of proceeding, but rather so compelling a critique of the exclusionary structures of modernism that have suppressed the subjectivity and sexuality of disabled people that a call for radical transformation is the only adequate response. Where in the early years of political activism and scholarship the notion that material changes in the social organisation of disability – simplistically named as progress – might settle the claims of those affected by discrimination and devaluation, the sustained challenge effected by postconventional perspectives to the organising binaries of western thought, and to the normativities of the subject and of sexuality, impel a far more complex approach that takes account equally of psychic, socio-cultural, and political domains. And to reconfigure the meaning of disability is indeed to enter into a dangerous discourse whose implications cannot be contained in a single constellation. Critical disability theory has the capacity to not just change the lives of a significant minority of people who are categorised as disabled, but to disrupt the whole nature of the relationship between differently embodied subjects. Following Deleuze in particular, the emphasis shifts from supposedly discrete and bounded categories to the dynamic of assemblages in which the disabled body is merely one component. If interdependency in the guise of connectivity were seen as simply a condition of becoming for all of us, then the dominant discourses of rights, choice, and self-determination could give way to a more open and productive model that did not need to focus exclusively on the putative benefits of civil status.

One key theme throughout the book that has come up time and again is the question of governmentality in its multiple guises. I have asked whether the demand for acknowledgement of the hitherto disavowed subjectivity and sexuality of disabled people might not deliver those who so identify to the deadening coils of a normativity that stifles creative flourishing. My wariness – but not quite rejection – with regard equally to the explanatory rigour of Freudian and Lacanian psychoanalysis, to the Foucauldian cycle of power and resistance, and to neoliberal legal notions of citizenship and justice, is mobilised by just such a concern. Each discourse offers a seductive vision of constructive possibilities but cannot avoid recourse to the establishment of a new status quo. Critical disability theory demands more, and although a Deleuzian approach has some positive openings, there is much more work to be done to follow through on the potential. The necessity is not to settle on a singular perspective, however, but rather to continue a process of intersectional exploration that is not afraid to utilise critique even in the absence of an alternative way forward. In his discussion of critique, Foucault is adamant that for it to be truly transformational – to be more than a simple readjustment of normativities – one must think differently and then 'transformation becomes both very urgent, very difficult, and quite possible' (1988b: 155). Though a reductive appraisal of the terrain might enhance the use value of any new reading, we cannot simplify questions of disability, and I would argue that what is needed is precisely a more sophisticated approach that refuses to flatten out the multiple layers of significance and meaning. As Derrida puts it, 'one shouldn't complicate things for the pleasure of complicating, but one should also never simplify or pretend to be sure of such simplicity where there is none' (1988: 119). The risk and promise of postmodernist modes of analysis are closely linked in the provisionality and fluidity that is the *sine qua non* of that approach. In jettisoning the anchors of modernist thought and all that depends on it, the inevitable loss of individual and socio-political certainty is offset by what is enabled: the capacity to move beyond existing categories of knowledge.

The promise and possibilities of postmodernism must always be tempered, however, by a continued awareness that there are other elements emerging in the era of postmodernity that may prove inimical to the well Being of disabled people. The development of gene therapy is a possible case in point and while I do not want to subscribe to any blanket condemnation of pre-natal genetic diagnosis, for example, there is a clear potential for new bioscientific practices that materialise the fundamental psychic disavowal of disability. That is precisely why I have put so much stress on the necessity for change in the cultural imaginary, for although rights and legal restrictions may provide some protection, both are commonly rescinded or overridden when they no longer serve the purposes of the state. Given, however, the pace at which the apparent certainties of the modernist world-view are being transformed both globally, where narratives of national progress and social order are challenged both theoretically and materially, and at the level of the individual where identity is destabilised, there is every opportunity to take a relatively optimistic approach. The contingency and flux that are characteristic of contemporary society at a macro-level are paralleled most clearly in the micro-politics of the body, and not least in the recognition that the putative wholeness, integrity, and distinction of the embodied subject is a phantasmatic structure that is increasingly contested. In other words, it is not simply that some exceptional – paradigmatically disabled – bodies fall short of the normative standard, but that embodiment in general is disordered and uncertain. Such an acknowledgement does not cover over difference, nor deny the specificity of the phenomenology of disability; rather it figures instability as the *unexceptional* condition of all corporeality. The inescapable implication is that instead of perceiving 'disabled' embodiment as an anomaly relative to some ideal, and as failing to qualify the person thus embodied for the full attributes of subjectivity, we should revalue it as just another variant on the infinite modes of becoming. Once corporeal integrity is shown to be a contested and contingent form of embodiment, rather than a privileged standard, then the realisation of diverse styles of subjectivity and sexuality in the disabled body is no longer contentious.

From the perspective of postmodernity, where radical and unpredictable transformations are already affecting each one of us, the modernist anxieties provoked by non-normative embodiment speak more to an imaginary and now lost past, than to future probabilities. The centrality of humanist ideals and of a defined human form itself – in whatever way that might manifest – are deeply contested not only by emerging theory but by developments in biotechnology. Even accepting the uncertainty of how such transformations might unfold, it becomes increasingly implausible that in the era of the posthuman, differences in corporeality could serve as any justificatory ground for exclusion and oppression. Of course nostalgia persists, but

I would suggest that an ethics based on the distinction between bodies is inadequate, and that a better model would at very least encompass a phenomenological approach that entails an ethics of contact and touch, and – more radically – an attention to Deleuzian assemblages and flows of connectivity. It is in the deconstructive work that those and other postconventional perspectives entail that the term set out in the 'Introduction' – dis/abled – reveals its significance. It connotes neither the one nor the other, but expresses instead a refusal to fall in with the normative pattern of binary structure, whether of material embodiment as such, or of the diverse attributes said to adhere to particular bodies. If dis/ability figures a space in which distinctions disappear, then the bioethical implications for those who are non-disabled and disabled alike are inescapable. As the unmarked category, the taken-for-granted standard, and the dominant power, it is, without doubt, the former who have the most ethical work to do. All the vulnerability – figuring putative and abject failures, lacks, and weaknesses – that has been thrown on to the others must be owned as a condition of becoming shared by all. As such, vulnerability is not a negative attribute but simply an expression of the contingency and incompletion that characterises all life. Where, in both history and contemporary society, disabled people have been made to bear the mark of instability and disorder, what postmodernism makes clear is that they are not exceptions to a general rule of security and predictability but merely ordinary exemplars of the process of living.

To reconfigure the meaning of disability alone will, however, not suffice, for – as the exploration conducted in this text demonstrates – it is part of a fluid nexus that depends on a parallel deconstruction of the notions of subjectivity and sexuality, particularly in the mode of desire. The dangerous discourses that I set out to investigate have proved inextricably linked together, and have opened up not just new ways of acknowledging and accessing either the subjectivity or sexuality of disabled people, but have placed a firm emphasis on the mode of becoming. The notion of becoming, which I introduced as an element in Merleau-Ponty's work, becomes with Deleuze increasingly transformative in its shattering of the sovereign subject, as well as always implying a working through of desire. The Deleuzian philosophy of what is termed 'becoming other/imperceptible' (Deleuze and Guattari 1987) intends a scenario in which the differential 'being' of any subject is always in a process of unravelling, largely through an acknowledgement of the multiple webs of connections that constitute becoming. In place of any predetermined and relatively unchanging subject as the holder of rights, life is defined not by an essential form or teleology, but by the energy, speed, and fluidity with which it enters into and sustains machinic assemblages. When Deleuze insists on the rhizomatic nature of life – by which he means its proliferation in ever-new forms along multiple and unpredictable pathways – he decisively breaks with the notion of an atomistic and corporeal subject, and signals a state of becoming that goes beyond any individual lived experience. To be able to say 'I' is a contingent project which, although it is actualised in an individual body that can represent personal value, is simply one part of the cycle of becoming that extends beyond the human and the singular body to figure a non-temporal and unstructured coalescence of creative forces. Rather than providing a stable centre, a reference point for agentic actions that reach out to and withdraw from others, the body is rather a conduit for the intensity of apersonal possibility and desire. In the unlimited mode of an assemblage, no single unit – be it a body, an action, an identity – has meaning or significance other than through its irreducible connections with other affects and flows of energy. It is a continuous process of transmutation, of becoming other that mocks the modernist obsession with boundaries and limits.

In taking up, and to some extent wilfully reshaping, a Deleuzian mode of thought, a radical transformation in the meaning of disability becomes possible, but as I have indicated from the start, it is a move with far-reaching consequences: perhaps it will change everything. What is clear is that the notion of dis/ability demands a reconfiguration of all our socio-cultural, political, and ethical considerations, and holds out the promise of crucially disrupting the existing imaginary. The deeply transgressive nature of the Deleuzian approach will perhaps appear too dangerous to many, but it is difficult to deny that the move must be towards transformation, not accommodation. Where the call for the liberation of disabled people from the hostile normativities of contemporary western society will, I believe, focus increasingly on the restrictions and invalidation imposed on sexuality, the point surely is not to rework existing definitions, but to look for something different, to take the risk of trying out the untested and unexpected. As Foucault (1988a) understands it, the first problem is to ask what sexual pleasure is, and what could be meant by the erotic, or by passionate relationship. If we are to avoid new forms of governmentality – if that is possible – then the radical transformation of all that is taken for granted is essential. Once we begin to think the body, the subject, and sexuality beyond limits, the consequent opening up of possibilities might seem too great to incorporate into the materiality of life, but once more it is necessary to remember that the first stage is to think differently. What follows on from that cannot be determined in advance. The legal theorist Isabel Karpin, who has worked widely on disability issues, offers a productive way forward in positing a *'norm of transgressivity*... [which takes] as its base unit a subject that is inevitably connected, vulnerable and dependent' (2005: 197, my emphasis). Karpin is fully aware of the ironic implausibility of the phrase, but insists equally that it refers to 'a standard state of being'. Karpin is in fact discussing the law, but her proposal can encompass all the discourses in which disabled people are invalidated. In any case the point of critique, however deconstructive it may be, is not to absolutely prohibit the use of what is contested, but to put it into circulation with its challengers where each is kept open to productive distortion and evolution. Like Derrida, Deleuze does not offer a practical programme for change, but rather a call to begin exploring other ways of becoming along new – even impossible – lines. Both provide a horizon of aspiration that is essential if dis/ability is to emerge from the rhetorical politics of identity, the claim on sociopolitical rights, and the fear that the necessity of making immediate life more equable for disabled people precludes the queer space of thinking outrageously and otherwise.

It is not my claim that either queer or Deleuzian theory will come to dominate critical disability studies, but rather that, vis-à-vis the academy, disability studies itself has a theoretical significance which, by working at the intersection of both, will be capable of taking on the role of critique that queer and Deleuzian theory now occupy. And, moreover, its abstract concerns will, I believe, progressively spill over into more substantive matters. Like feminism it already has an effective activist base, and just as feminist themes and theory have come to necessarily infiltrate every aspect of thinking about the body, about the imaginary, about the law and social policy, about psychoanalysis and the Symbolic, so too dis/ability demands to be taken into account. The theorisation of dis/ability decisively extends, distorts, deconstructs, and – yes – partially incorporates its own antecedents to open up innovative and creative ways of thinking about the attributes and yet to be realised capacities of differential embodiment, just at the point at which the singular subject gives way to the uncertainty of multiple becomings. It is a moment that undoubtedly complicates all our prior assumptions about boundaries and identities, whether individual, local or global, and that puts into question the putative security and familiarity of a normativity that has been both oppositional and desired. To engage with a dangerous discourse is always risky, but that should mean proceeding with some caution, not avoiding the discomfort altogether. It is not simply that the risk is worth taking for the adventure of reconfiguring disability to feel and act differently, but that it is an ethical necessity – a responsibility to otherness – that leaves no-one behind.

Notes

Introduction

1. The terminology around disability is highly contested, not least around the nomination of those who putatively occupy the category. The current preference within both critical disability studies, and some but no means all activists circles, is for 'disabled people' rather than 'people with disabilities'. After a recent history in which the latter was promoted as a reaction to older and more evidently stigmatising terms such as handicapped, retarded, cripples and so on, or to supposedly more positive alternatives such as differently abled, physically challenged or special needs, the use of so-called *people-first* language forms is now seen as a confirmation of the person that fails to encompass the significance of disability (Overboe 1999; Titchkosky 2006). It is as though disability were a contingent add-on rather than a fundamental element of identity. Nonetheless 'people with disabilities' is in very widespread use among disabled and non-disabled people alike, and indeed some academic journals mandate it. On either side such policing is worrying, and although I prefer 'disabled person' as more adequately denoting the process of embodiment, I shall not entirely exclude its alternative where it might seem contextually more appropriate.

2. From the Aristotelian notion of women as deformed men, through the medieval and early modern idea that geographically remote races were quite literally monstrously embodied, to recent much-disputed DSM classifications of gender dismorphia, and the current rejection of the label of disability by the Deaf community, there is no unified, singular or unchanging model.

3. Deborah Marks (1999) suggests the term CAB – contingently able-bodied – rather than TAB would better capture the uncertainty that surrounds disability.

4. As Foucault's work makes clear, discourse itself intends not simply a structure of language but a material practice.

5. For earlier explorations of the notion of intercorporeality, see my work with Janet Price (Price and Shildrick 1998; Shildrick and Price 2001) in which our collaboration was to a significant extent a, sometimes playful, sometimes necessary, performative transgression of the usual ablebodied/disabled binary.

1   Corporealities

1. Without using Kristeva as a theoretical touch point, Deborah Marks (1999) is nonetheless one of the few theorists who not only insists on the irreducibly

psychic dimensions of disability, but enquires into the operation of anxiety, particularly from the perspective of the one who considers herself
non-disabled.

2. The classic medical presentation of the disabled body – and more generally of any morphological anomaly – entails an absolute division between
the knowing subject and known object which *displays* its difference from a
distance. I am reminded of the shock engendered by Princess Diana when
she broke the bounds of medical propriety – albeit in a context in which it
was well understood that the HIV-AIDS was not transmitted by skin to skin
contact – by touching patients in an AIDS facility.

3. That openness of course is precisely the rationale given for the effectiveness of 'Eastern' based biomedical procedures such as reiki and shiatsu which
rely on some level of corporeal interchange between practitioner and client
(Paterson 2005).

4. See Chapter 7.

5. Sexuality is the paradigmatic site in which it is clear that corporeal performitivity is inherently socially inscribed, always engaged with the other, and
never singular as Butler sometimes seems to imply. As Ros Diprose (2002)
points out, my meanings are never constituted personally.

6. For further development of this theme, see Shildrick 2002.

7. The theme is fully developed in Chapter 4, 'Subjectivity, sexuality and
anxiety'.

8. We have published several joint papers, but those most relevant to
this chapter are Price and Shildrick 1998 and 2001 both of which
have used autobiographical material to theorise aspects of disability,
like sensation and emotion, that more conventional approaches have
discounted.

9. It is easy to imagine, for example, Miho Iwakuma's speculation on the extension of embodied subjectivity in the arena of sexuality – 'when a lover of a
person in a wheelchair touches the chair, s/he shivers as if the flesh of the
person were caressed. In a sense, the person *was* touched…' (2002: 78) –
being dismissed as fetishistic.

10. Meekosha (2004) provides an excellent account of the differential genealogy
of UK and US models of disability that has mobilised the SMD in the former
and a turn to questions of identity and experience in the latter. See also
Marks (1999).

11. The irony is that despite the strength of feminist scholarship on the
body, there has not been any widely promoted uptake of disability as
another of the differences – alongside sexual difference – that would
enhance the contestation of normative and masculinist standards that
would devalue embodiment. It still appears that the overwhelming majority of texts that interrogate the interface of feminism and disability are
written by women who are themselves disabled in some degree. See, for
example, Begum (1992), Morris (1996), Wendell (1997), Garland-Thomson
(2002), as well as Meekosha (1998) and Corker (2001). Far from display-
ing an appropriate sensitivity towards those who would speak from their
own experience, the omission, I believe, both speaks to a dereliction of
responsibility and fails to follow through on the implications of our own
feminist scholarship.

2   Genealogies

1. Clearly the bodies that link us with our ancestors greatly differ from contemporary norms in terms of size, weight, musculature, and longevity, while
   the development of prosthetic enhancements have had equal impact on
   capacity. From the simplest crutch, through the use of spectacles, to the
   most technologically advanced bionic limb, the body has rarely been materially unmediated. The mistake, however, would be to assume that changes
   represent a continuous and unproblematic curve of progress.
2. See, for example, the issue of *Cultural Studies* devoted to 'Genealogies of
   Disability' (Diedrich 2005), the edited collection, *Foucault and the Governmentality of Disability* (Tremain 2005), and a number of individual essays in
   Disability/Postmodernism (Corker and Shakespeare 2002).
3. See my account in *Embodying the Monster* (Shildrick 2002) for the ways in
   which bodily anomalies could be read as marvels that signalled both natural
   fecundity and the unlimited creative power of the godhead.
4. The social model of disability (SMD) refers throughout to the British-derived
   model developed by Michael Oliver (1990) in particular, and not to the
   rather different, and less-overtly politicised, US social model. While the first
   sees socio-political organisation as producing disability, the second is more
   personalist and understands disability to result from discrimination directed
   against certain differently embodied people who are unable to fit normative role models. As David Pfeiffer explains: 'Researchers using the UK Social
   Model will analyse social structures and their impact on people with disabilities. Researchers using the US Social Model will analyse social roles and
   attitudes toward failure to fulfill them' (2002: 234).
5. Moira Gatens characterises the cultural imaginary as, 'those ready-made
   images and symbols through which we make sense of social bodies, and
   which determine, in part, their value, their status, and what will be deemed
   their appropriate treatment' (1996: viii). The imaginary is the unifying locus
   of all the categorical, corporeal, and socio-political demarcations that hold
   apart what is to count as normal, and what abnormal.
6. See *Embodying the Monster* (Shildrick 2002), for a more detailed discussion of
   this point with reference to the history of the monstrous.
7. In a recent article, Stiker (2007) has acknowledged the extensive influence of
   Foucault – and particularly *Madness and Civilization* (Foucault 1965), and the
   notion of *biopower* – on his own work.
8. In a related way, a whole range of everyday experiences common to
   women – such as menstruation or pregnancy – were deemed equally dangerous and polluting. Indeed, Julia Kristeva claims that the force of biblical impurity is 'rooted...in the cathexis of maternal function – mother,
   women, reproduction' (1982: 91). The common denominator across a range
   of defilements seems to be that the ideal of corporeal unity has been
   breached.
9. Stiker, as always, teases out the contradiction. In discussing the new system, he asks: 'But is misfortune any the more averted?....A completely
   new form of integration – and of exclusion – has been introduced' (1999:
   35–6). And he goes on to assert that the demise of religiously sanctioned
   prohibition – which paradoxically demanded *social* responsibility towards

those excluded – throws the weight of response on to the individual: 'our relationship to disability … depends entirely on ourselves' (*ibid*). As he sees it, it is charity alone that must negotiate the encounter.

10. For a full discussion of this highly complex debate, see 'No Monsters at the Resurrection' (Pender 1996).

11. Where I perceive a necessary blurring of categories, Park and Daston (1981), discussing similar material, attribute clear distinctions between natural, supranatural and supernatural to the relevant texts.

12. An enormous proliferation of such texts included Boaistuau (1560), Fenton (1569), Aldrovandi (1642), and Buffon (1749), as well as travellers' guides warning of the strangely embodied quasi-human races to be found at the margins of the known world. The most famous is *Mandeville's Travels* (1967) which first appeared sometime in the fifteenth century.

13. Stiker cites the historian Philippe Ariès as one influential source of this view which he sums up thus: 'Normality was a hodge-podge, and no-one was concerned with segregation, for it was only natural that should be malformations' (Stiker 1999: 65). In his own discussion of a relative lack of material concerning disability coming from the Middle Ages and early Renaissance, Stiker is more circumspect and introduces copious evidence of the marginalisation of disabled people, and adds: 'Perhaps the Middle Ages did not speak of the disabled lest it awaken too many terrors' (1999: 70).

14. Paré's ambition is all the more remarkable in the light of François Jacob's claim that his age was one in which '(n)o distinction was made between the necessity of phenomena and the contingency of events' (Jacob 1989: 19).

15. The decline in the appeal of monster books as such does not imply any lessening of psychic unease and attraction. Modernist scepticism with regard to the 'fantastic' representations of corporeal anomaly is always highly tempered, as is evident in texts such as the lurid bioscientific accounts given in Gould and Pyke's *Anomalies and Curiosities of Medicine* (1897) or Leslie Fiedler's scholarly but lavishly illustrated *Freaks* (1978).

16. To say that the experience of anxiety is transhistorical is not to deny the specificity of the forms that it may take.

17. The term appears to have originated in a German treatise – translated as *Permitting the Destruction of Unworthy Life* – written in 1920 by a psychiatrist and a lawyer, Alfred Hoche and Karl Binding respectively, who laid the grounds for involuntary euthanasia in the interests of the state. By the late 1930s, the German Reich had already sanctioned the destruction of disabled infants, and in wartime moved on to the infamous *Aktion T4* programme that led to the murder of many thousands of disabled adults (see Friedlander 1995; Mostert 2002; and Armer 2005 for further details). Despite some overt opposition, and the relative secrecy and euphemistic cover for such actions, claims that they went against the broad views of lay citizens are difficult to sustain. As Mostert remarks: '(R)equests for mercy deaths were increasing and were generally viewed as more acceptable, whether conducted by individual citizens or the state. Essentially, disability was widely acknowledged to be a legitimate justification for murder' (2002: 160). Widespread public opposition was slow to develop, and although the official killing centres of *Aktion T4* closed in late 1941, hospital- and institution-based euthanasia continued throughout the war (Mostert 2002).

18. Gallagher (1990) indicates that the eugenic abuses against disabled people were never recognised as war crimes and played no part in the post-war trials of prominent Nazis. See also Mitchell and Snyder (2003).

19. Brady *et al* starkly make the point in their Australian study in which they claim sterilisation was 'related primarily to two characteristics – gender and disability' (2001: ¶2).

20. The connection between dehumanisation and violence is somewhat ambiguous, and I do not want to imply a simple cause–effect relation. Jonathan Glover (2000) has written extensively about various examples of large-scale state-led atrocities that deploy the tactic of dehumanisation, though without, I believe, referencing the National Socialist assault on disabled people.

21. Foucault (1965) makes a similar point in his contention that the strategy of what he calls the 'great confinement' – that catch-all institutionalisation of both physically and mentally disabled people during the Classical age – was abandoned finally in favour of new distinctions between physical and mental anomalies, that figured inclusion and surveillance on the one hand, and continued segregation and disciplinary control on the other.

22. See Ott, Serlin and Minh (2002) for an analysis of the rise of prosthetics. Serlin explicitly makes the link with normative sexuality when he refers to wartime practices as 'the fiercely heterosexual culture of rehabilitation medicine, especially its orthodox zeal to preserve the masculine status of disabled veterans' (2006: 170).

3   Contested Pleasures and Governmentality

1. To speak of ethics in a postconventional sense is to move beyond the parameters of rights and duties, or of harms and benefits, in order to engage with the nature of encounter, and the relation between self and other.

2. See Shildrick and Price (1996) and Tremain (2001, 2005) for a Foucauldian analysis of disability practices.

3. Both modes – repression and silencing – come together in Keywood's analysis of two recent English cases in which she contests the assumption of gender bias in the pertinent law, but goes on to claim that law constitutes the learning disabled subject 'as being of no "sex"' (2001: 186).

4. I am thinking of those who may be unable to experience genital sensation or to verbally communicate their needs and desires; of women who cannot conceive or carry a foetus to term; or of lovers whose sexual activity requires the physical support of third parties.

5. See, for example, White (2003) and Job (2004). In her small qualitative study of professional caregivers employed under independent living arrangements, Earle notes: 'The perception of the severity of an individual's impairment seemed to be associated, not necessarily with asexuality, but with an incorrect belief that it habitually relates to a physical inability to engage in sexual activity' (1999: 317).

6. Disabled people now appear frequently on television, for example, both as commentators and reporters, and as fictional characters who may appear in a variety of respected occupations and locations, but far less commonly as engaged in sexual encounters. The implicit message conveyed by such images seems to be that in the wider sense, disability makes no difference

and that we are all the same under the skin. The media take scant account of the differential experiences – often including ones of a sexual nature – of those people whose very forms of embodiment preclude assimilation. In any case, as Henri-Jacques Stiker has argued of disabled people, 'they are spoken in order to be silenced' (1999: 134).

7. All these scenarios are referenced in detail in Shildrick (2004a). The 2005 Sex Survey organised by Simon Parritt on behalf of the UK newspaper *Disability Now* found plentiful evidence among its 1115 respondents of sexual activity and concerns, but as Parritt himself remarked in summation: '(A)lmost all charities and prominent disabled people shy away from addressing sexual and relationship inclusion. This is a denial and de-sexualising of us by our own organizations' (2005: ¶4).

8. One of the most powerful recent images of disability has been Mark Quinn's large-scale statue, *Alison Lapper Pregnant*, which was on temporary display in Trafalgar Square, London during 2005–2007. The artwork unashamedly alludes to disability and sexuality – Lapper's naked body leaves the viewer in no doubt – but nonetheless once again limits sexuality to the mode of reproduction. The public reception of the work has been broadly favourable, but many respondents have remarked on Alison Lapper's status as a pregnant woman among the dominant imagery of martial masculinity rather than on her very evident disability.

9. This term refers to a specific state allowance available in the United Kingdom to a limited number of disabled people. Funds from similar charity-sponsored schemes may also be obtained, often as a supplement to state provision.

10. The claims of disability activists and scholars that independent living provisions can enhance the expression of sexuality for disabled people have appeared for some years (see, for example, Ratzka 2002; Shue and Flores 2002). The effectiveness of independent living has recently been endorsed by the work of the UK state-sponsored *Office of Disability Issues*. Reviewing its work during the first year of operation, Ben Furner has commented: 'Independent living issues take pride of place, with promising results from a pilot study of schemes designed to give disabled people more choice and more control over what they do and how they do it' (2006: 1).

11. Although heteronormativity may generate some recognition of sexual expression as a basic need, there are, as Richardson argues, 'certain social groups where the assumption of sexuality as a universal human "need" appears to be challenged. Stereotypes of disability, for example, include assumptions of asexuality; of lack of sexual potential' (2000: 109). In the *Disability Now* Sex Survey (What YOU said in the survey, 2005), the level of self-declared interest in sexuality among disabled people was tempered by a recognition of the social attitudes that constrain expression. As one anonymous respondent wrote: 'I feel my sexual needs are completely invisible....no health care professional has ever asked me about my sexual life even though I have been seriously ill for nearly 17 years' (2005: ¶12).

12. I make no attempt here to enquire into jurisdictions other than the United Kingdom, except to note that the level of legal intervention into the conjunction of disability and sexuality may vary in accordance with the parameters of publicly acceptable sex(uality) in general.

13. Unlike most of the United States, where both the buying or selling of sexual services is illegal in itself, and at least treated as a misdemeanor, UK law does not criminalise sex workers *per se*, nor their clients, and there is no law against procurement unless it is for gain. Nonetheless, the *Sexual Offences Act 2003* (Sections 17 and 21) makes it a criminal abuse of trust to facilitate sexual acts for those under the age of 18 in institutional care, even though the age of consent in the United Kingdom is otherwise 16 years. Section 31 outlines a similar offence with regard to those classified as mentally handicapped, whatever their age. And in addition, The *Sexual Offences Act 1967* (Section 4) lists any third party procurement of homosexual acts as an offence.

14. These issues may be particularly difficult for some feminists to negotiate because they are so focused on a masculinist appropriation of women's bodies. All the general objections to and defences of sexual exploitation have resonance here, but with the proviso that the power relations that are being played out by no means fit the normative pattern. It is not that I think there is nothing to debate, but rather that the whole question has been obscured by issues of law and public policy.

15. For the sake of mutual understanding, the ideal contract of employment would stipulate such matters, but I can find no evidence that such overtly worded contracts exist.

16. In my own informal and non-directive discussions with both sides about the relationship between disabled women and their assistants, my highly anecdotal impression is that although it is now perfectly straightforward to speak in confidence about, for example, facilitated illegal drug use, the issue of facilitated sex often remains off limits.

17. Despite a title – 'My Home, Your Workplace: People With Physical Disability Negotiate Their Sexual Health Without Crossing Professional Boundaries' – that includes the phrase *sexual health*, which, as Browne and Russell (2005) themselves point out, usually refers to the presence or absence of disease, the article is actually concerned with what people with disabilities understand to constitute *sexual well-being*. For the purposes of the paper, it is broadly defined as 'the capacity to enjoy and control sexual behavior in line with a personal and social ethic' (2005: 376).

18. *Professional carers* is the writers' preferred term.

19. Those designated as others under conditions of heteronormativity are not limited, however, to sexual others. 'Improper' corporeal configuration is closely associated with 'improper' sexuality, as I have shown, to the extent that heteronormative society impels what McRuer (2006) calls compulsory ablebodiedness.

20. Two decades or so on, safe sex is being increasingly discarded in the gay male community in favour of more risky practices such as bare-back sex. The cycle of power/resistance moves on.

4   Sexuality, Subjectivity and Anxiety

1. Congenital disability is the limit case, but even acquired disability is often sufficient to block recognition of a sexuate past.

2. As well as morphological difference, differences in self-presentation that arise from developmental disabilities may also play a part here.
3. It is immediately apparent that a move away from the notion of a sovereign self and a turn to understanding self-becoming as a matter of intercorporeality and psychic intersubjectivity might suggest an unconscious *pull towards* other bodies. Even so, given that the possible origin of such a pull might lie in the deeply repressed corporeal intimacy of infancy, it would again – as infancy itself is forgotten – invite a regulatory covering over.
4. See, for example, Tepper (1999), Wilkinson (2002), and Sherry (2004). Despite a plethora of new work in the area of critical disability studies, however, the movement in general terms has been fairly sexually conservative, and, as I outlined in Chapter 3, has not often given *public* voice to alternative forms of sexuality.
5. In the interests of space, I shall not pursue the Kristevan notion of the abject further here although it undoubtedly provides insights in the general 'untouchability' of the disabled body (Shildrick 2002), and could be applied more specifically to the domain of sexuality.
6. In addition to the authors mentioned in the text, Deborah Marks (1999) has paved the way for a wider acceptance of the benefits of a psychoanalytic approach. While she strongly endorses the social model of disability for its rejection of a pathologised past, Marks, nevertheless, breaks new ground in insisting on the importance of psychic dimensions. Brian Watermeyer's more recent essay 'Disability and Psychoanalysis' (2006) is strongly indebted to her, whilst pushing the argument further. In particular he develops the notion that pity towards, and an idealisation of, disabled people covers over and disavows an unconscious sadism towards disability. Once again it is the anxiety of the non-disabled majority that is at stake.
7. As Wilton (2003) suggests, it might be instructive to trace the extent to which the wound of castration is psychically figured by the disabled body. Certainly Freud insists that the uncanny – which I have evoked to figure the disabled body – is always linked to castration.
8. I am suggesting that all adult sexuality is affected by such a desire, but it is clear that the individual's specific experience of being mothered may further exacerbate the nostalgia for the dependent body. It is highly plausible that the attraction of fragmentation might be heightened in instances where the care-giver's ministrations had been especially attentive and seductive during periods of illness or trauma that could have quite literally threatened the infant's corporeal integrity.
9. The term 'other' in this context is itself paradoxical insofar as the maternal-infant dyad – and equivalent figurations – is at once irreducible and torn apart.
10. I do not mean to imply that the person with a disability *should* be equated with her sexuality, but rather that just such a synedochal identification does in fact take place.
11. Amputee devotees, who are most likely to be non-disabled themselves, may both desire those with impairments, and in rarer cases wish to become them by excising their own limbs. It is not clear, as Nikki Sullivan (2008) notes, that the latter condition – despite being classed as a paraphilia – is in fact sexual in nature, but may represent instead an instance of identity disorder.

It does fall, however, within an expanded notion of desire, such as that posited by Deleuze and Guattari.

12. In his passing discussion of the fetishisation of disabled bodies, Robert Wilton notes, 'the origin of this process lies not with the fact of physiological difference, but with the ultimately untenable character of the phallic (non-disabled) subject' (2003: 380). See also Aguilera (2000) and Kafer (2004) on the controversial and substantive issue of devotees.

13. My thanks to Kay Inckle of Trinity College Dublin for initially raising this point with me. I am reminded too of Luce Irigaray's titanic struggle with Freudian/Lacanian psychoanalysis on behalf of the feminine. Although there seems to be some celebration – in essays such as 'La Mysterique' (1985) – of the freedom of a feminine *jouissance* excluded from the Symbolic, Irigaray's more consistent position is that the Symbolic must be reconfigured to accommodate, and give voice to, its others. Regardless of my doubts about sexual citizenship, I have no hesitation about the desirability of sexual subjectivity.

14. In a similar way, it is that condition of the subject that situates corporeal anxiety as indivisible from ontological anxiety. I take it as a given that the subject is always embodied, and that the 'I', in the sense of the ego, is, as Grosz puts it, 'both a map of the body's surface and a reflection of the image of the other's body' (1994a: 38).

15. For a more optimistic account that specifically rereads the potential of the Freudian/Lacanian model in the light of queer sexuality, and the significance of HIV/AIDS, see Tim Dean (2000). Dean writes: 'without an appreciation of the unconscious, queer sexualities themselves become normalizing (paradoxical though that sounds), insofar as sexuality becomes wedded to identity' (2000: 6). It may be that an analogous move could be made with regard to disability.

16. Even those advocating polymorphous pleasure may appear locked into the conjunction. In her commentary on the 1996 film *Crash*, directed by David Cronenberg, Linda Williams argues that it is 'genuinely polymorphous perverse' (1999: 43). The narrative of *Crash* specifically links sexual desire to the broken bodies of a group of men and women who deliberately seek out self-injury through enacting car crashes. The connection between sex and death is a long-established psychoanalytic trope, and in comparing *Crash* with the documentary, *Sick* (Kirby Dick 1997), on the life work and death of disabled performance artist and self-declared sado-masochist Bob Flanagan, Tom Shakespeare speaks positively of 'a radical choice to embrace and explore injury, mutilation and mortality' (2005: 59), which raises 'uncomfortable questions about our secret desires' (62).

5 Transgressing the Law

1. It is not for nothing that Derrida's most influential contestation of legal discourse is called 'Force of Law: The "Mystical Foundation of Authority"' (1990).

2. See also *Embodying the Monster* (Shildrick 2002), Chapter 1, for additional development of the contemporary debate.

3. The principal source used by Foucault is *Traité d'embryologie sacrée* (1745) written by the Italian priest, F. E. Cangiamila.

4. All examples taken from *Embodying the Monster* (Shildrick 2002: 19, 34).

5. Apart from the lectures series, *The Abnormal*, Foucault scarcely mentions disability directly, but his wider work on bodies, on power/knowledge, and on governmentality is saturated with relevance. Nor should it be supposed that Stiker merely follows Foucault. *The History of Disability* (1999) is undoubtedly influenced by Foucault, but largely develops its own trajectory.

6. See Shildrick and Price (1996) for an analysis of how one social security benefit directed to those with disabilities acts as an instrument of both corporeal construction and regulation.

7. The debate over the efficacy and implications of 'neutral' universal rights versus 'special' rights is well-developed, and beyond the limitations of this chapter. In most cases the privileged term of the problematic of sameness and difference as the competing grounds of rights discourse is either gender or race (see Mackinnon 1987; Cornell 1991, 1995), but it can, with due caution, be transposed to the question of (dis)ability. For an interesting run-through on one aspect of the sameness/difference conundrum with respect to both physical and developmental disability, see Goering (2002).

8. I am acutely aware of the dangers of such slides between the differing modalities of power that constitute and embody us as variously sexed, gendered, raced, classed, and endowed with differential physical and mental abilities. But while it is rarely adequate to effect a direct substitution of one category for another, there are certain significant similarities in the operation of power that justify extending the empirical limits of applicability. When Foucault speaks of the disciplinary nature of power/knowledge, for example, his focus is on the production of sexualities, but finds equal resonance in reference to other dimensions of bodily comportment such as (dis)ability or gender. A related and equally important concern is to stress – as does Wendy Brown in deploring the compartmentalism of much antidiscrimination legislation – that the various categories of identification that I have mentioned, and others, are never discrete, or even simply overlapping, entities, but inextricably intertwined in the very moment of production. As Brown summarises: 'we are not simply oppressed but produced through these discourses...through complex and often fragmented histories in which multiple social powers are regulated through and against one another' (2002: 427).

9. It is not an assumption that I share, or one that even addresses what I take to be the pertinent issues (Shildrick 2002).

10. The heart and lungs of 'Mary' were underdeveloped to the extent that those of 'Jodie' supplied all the oxygenated blood for the entity Jodie–Mary through a shared aorta. Medical prognosis indicated that without separation, Jodie's heart would fail within a few months, and with separation Mary would be denied the oxygenated blood necessary to her survival.

11. See Michalowski (2001) for an entirely normative discussion of the Court of Appeal deliberations on the issue of consent.

12. Exceptions include Sheldon and Wilkinson (1997), Hewson (2001), and Munro (2001), who variously suggest that the autonomous individual may

be an inadequate starting point in difficult cases. None, however, moves from an appreciation of the implications of the one plus one of connection and relationality to a more radical acceptance of the irreducibility of a transgressive intraconnectivity.

13. There is not scope within this chapter to further explore the general discursive instability of law, but the 'infelicities' that Ziarek notes have wider application. Given the psychoanalytic approach to juridical discourse that Goodrich (1995) suggests, one might apply a Freudian analysis to such puzzling slippages.

14. Compare this circumspection with the actual operation of the Court of Appeal in the Jodie–Mary case. Whilst admitting the difficulty of finding precedents, Ward L. J. asserts that although in broadly parallel cases, there 'are always anxious decisions to make ... they are invariably eventually made with the conviction that there is only one right answer and that the court has given it'; and he goes on to observe: 'This court is a court of law, not of morals, and our task has been to find, and our duty is then to apply, the relevant principles of law to the situation before us – a situation which is quite unique' (*Re A* 2001: 6).

15. In common with so many other theorists, it does not appear that Derrida has directly addressed disability. Nonetheless, his remarks with reference to abortion, or bio-engineering, for example, leave little doubt that he would have figured the relation of the disabled body to the law as deeply unsettling, not least in its intrinsic exposure of the limitations of the legal subject.

16. What Derrida intends here is not only a hospitality that is unconstrained by pre-existing rules and structures (as I go on to discuss), but one in which reciprocity plays no part. Such hospitality is given openly without expectation of return, much less of self-accruing benefit.

6   Queer Pleasures

1. David Serlin notes: 'Disability studies and queer studies are ... allies in their shared commitment to demystifying the cultural and political roots of terms like *normal* and *healthy* and *whole* at the same time they seek to destigmatize the conceptual differences implied by those terms' (2006: 159).

2. *Abnormal* (Foucault 2003a) comes closest to exploring disability, albeit indirectly.

3. See Shildrick (2005b) for an extended discussion of the sexed/sexuate performitivity of conjoined twins.

4. It is perhaps somewhat unusual to move towards a DeleuzoGuattarian perspective through phenomenology, but whilst Deleuze eschews phenomenology as such, he clearly takes up the notion of reversibility, and other aspects of Merleau-Ponty's work. As Dorothea Olkowski remarks: 'desire turns upon Deleuze's conception of the body, so that when the connections between flesh, feeling, and desire are examined in terms of the body, the relationship between Merleau-Ponty, Bergson and Deleuze begins to reveal itself to be as deep as it is broad' (2002: 11).

5. Vivian Sobchack, for example, who has a prosthetic leg, remarks: 'in most situations, the prosthetic as lived in use is usually *transparent*; that is, it

is as "absent"…as is the rest of our body when we're focused outward to the world and successfully engaged in the various projects of our daily life' (2006: 22).

6. I do not want in any way to imply that prostheses are peculiar to those experiencing disability. On the contrary as Bernard Stiegler points out: 'The prosthesis is not a mere extension of the human body; it is the constitution of this body *qua* "human"' (1998: 152–3). Mitchell and Snyder make a similar point when they write, 'the prostheticized body is the rule, not the exception. All bodies are deficient in that materiality proves variable, vulnerable, and inscribable' (2000: 7). Nonetheless, it is disabled people who may be made the more conscious of the extension, substitution or supplementation of their bodies insofar as those modes are taken to speak to an originary *lack*.

7. I have been particularly impressed by the work of two graduate students at York University, Toronto. Loree Erickson (2007) has very effectively used photographic and video material to exhibit her own sexual inventiveness that highlights her bodily difference and long-term wheelchair use, while Jenn Paterson (2007) has made a preliminary study of the participation of disabled people in the Toronto BDSM (bondage, dominance, sadism, and masochism) community.

8. It is unlikely, however, that even this polymorphous pleasure would fully appeal to Deleuze and Guattari. As Parisi explains, their understanding is that the economy of pleasure 'represses all divergent flows from the cycle of accumulation and discharge, the imperative of the climax: the channelling of all relations towards the aim of self-satisfaction' (2004: 198–9). See also Deleuze (1994) for his disagreements with Foucault over the concept of pleasure.

9. Deleuze and Guattari pose the question of 'how to think about fragments whose sole relationship is sheer difference…without having recourse to any sort of original totality…or to a subsequent totality that may not yet have come about' (1984: 42).

10. It is worth noting that where the ideas developed by Deleuze and Guattari with regard to the connectivity and implications of desiring machines have struggled for understanding, the similar and almost cotemporaneous – albeit partially ironic – imaginings of Donna Haraway in 'A Cyborg Manifesto' – originally a 1983 conference paper – have become, for feminist and queer theorists at least, seminal fare. Of the 'illegitimate fusions of animal and machine' she writes: 'These are the couplings which make Man and Woman so problematic, subverting the structure of desire, the force imagined to generate language and gender, and so subverting the structure and modes of reproduction of "Western" identity, of nature and culture, of mirror and eye, slave and master, body and mind' (1991: 176).

11. Dan Goodley and Griet Roets (2008) have attempted something similar with regard to developmental disabilities.

12. I am grateful to Kelly Fritsch (2008), whose MA dissertation emphatically brought this work to my attention.

13. It is interesting that Overboe (2007b) also calls on the language of exception, but in a very different way to Prendergast (2008). He refers to Agamben's notion of bare life – that is life beyond the threshold of humanity, as much disability is perceived to be – as 'a state of exception' (Agamben 1998).

The counter for Overboe is precisely the Deleuzian-framed 'life-affirming expression that is not confined to binary thinking' (2007b: 29).

14. Disability theorists have approached the notion of queer in both more and less radical ways, but most would concur with Michael Warner that queer is defined 'against the normal rather than the heterosexual' (1993: xxvi). See in particular work by Shelley Tremain (2000), Robert McRuer (2003, 2006), McRuer and Abby Wilkerson (2003), as well as several other articles focusing on the intersections between disability and queer in a recent issue of *GLQ: A Journal of Lesbian and Gay Studies* 9.1–2 (2003). Many of the papers from the seminal 2002 *Queer-Disability Conference* in San Francisco are available online at www.disabilityhistory.org/dwa/queer/paper_bell.html

15. Although to date I know of few other disability theorists working directly on and around Deleuzian insights, this is, I hope, no mere rhetoric. In critical disability studies at least, the limits of liberal notions of equity and justice have already mobilised a willingness to explore alternative paradigms, and specifically queer theory. Given the material concerns at stake, the step into a more fully committed Deleuzian approach seems inevitable. To make the move in the reverse direction, to recognise that the field of disability might serve as a exemplar of many of the key concepts – not only linked to desire but to the notion of becoming-minoritarian – may take longer, but I take it that it is as apposite to Deleuzian theory as gay sex once was to the emergence of queer theory.

16. Although dis/ability is not mentioned, Eve Kosofsky Sedgwick offers an exemplary elucidation of what queer refers to: 'the open mesh of possibilities, gaps, overlaps, dissonances and resonances, lapses and excesses of meaning when the constituent elements of anyone's gender, of anyone's sexuality aren't made (or *can't be* made) to signify monolithically....At the same time, a lot of the most recent work around "queer" spins the term outward along dimensions that can't be subsumed under gender and sexuality.... "Queer" seems to hinge much more radically and explicitly on a person's undertaking particular, performative acts of experimental self-perception and filiation' (1994: 8–9).

7   Global Corporealities

* This chapter is based in part on work done together with Janet Price and published as 'Deleuzian Connections and Queer Corporealities: Shrinking Global Disability' (Shildrick and Price 2005–2006). The ideas arose from extended discussion and writing over several months that refined an original joint plenary presentation to the *Women's Studies Association* annual conference in 2004, but the form given here is my own development of that collaboration. It is impossible to quantify Janet Price's contribution to the wider process; suffice it to say that any acknowledgement would be insufficient.

1. Although disability rights movements are developing in many of the countries of the global South like India and Brazil, there continues to be a severely restricted range of writing around disability and globalisation. The literature that has emerged addresses issues both from the perspective of rights, as in Beresford and Holden (2000), Rioux and Carbert (2004), from the

economic point of view, as in Stienstra (2002) and Hurst (2003), or simply lists developments, as in Mayhew (2003).

2. Estimates suggest that figures for all categories of disability combined will rise to 1 billion by 2050, with a corresponding rise in the proportion of disabled people from around 8% to 14% (Mayhew 2003). Interestingly, that latter increase will impact largely on developed countries where the demographic is shifting to an ageing population.

3. Rosi Braidotti argues that queer theory is distinct from Deleuzean theory insofar as queering would be 'merely a proliferation of quantified differences and not a qualitative de-centering of hyper-individualism. Becomings are the sustainable shifts or changes undergone by nomadic subjects in their active resistance against being subsumed in the commodification of their own diversity' (2005–2006: ¶29). Her dismissal seems more appropriate to queer politics than queer theory.

4. See Wendy Brown (2002) for a detailed explication of the 'paradox of rights'. Hesitant disability theorists should note, however, that the critique of rights does not entail declaring that the pursuit of them is entirely redundant.

5. Explicit work on the links between queer theory and disability theory is as yet limited, and tends to address sexuality primarily. See Shelley Tremain (2000), but also Mark Sherry (2004) who makes much wider connections, and Robert McRuer and Abby Wilkerson (2003), both for their own introductory essay which touches on the issue of globalisation, and for the other essays in the 'Desiring Disability' issue of *GLQ* that signals the beginning of a strong contestation to mainstream commitments to oppositional identity politics.

6. Although it is somewhat unusual to associate Deleuze and Guattari with phenomenology as such, there is a familiar phenomenological thread running through their emphasis on the body as the locus of both unfolding affects and exterior influences.

7. See, for example, Harvey (1989), Giddens (1990), Hoogvelt (1997), and Bauman (1998).

8. Naomi Klein (2002) and the *Notes from Nowhere Collective* (2003), for example, have both expressed positive hopes of a global coming together. Alongside the option of resistance through the refusal of burdens, they recommend the embrace of what Hardt and Negri term, 'an alternative political organisation of global flows and exchanges' (2000: xv).

9. Power may refer of course to either *pouvoir* (power over) or *puissance* (potentiality/power to), though confusingly the *noun 'pouvoir'* seems to belong to the latter category. Although the relation between them is irreducible and fluctuating – and perhaps particularly so in the case of disability, as the life and death of Christopher Reeves (the disabled actor who had previously played Superman) illustrates – both are destabilised by a Foucauldian and/or phenomenological reading, at least insofar as intentionality is concerned. For Deleuze and Guattari, the rhizomatic nature of power entirely precludes individual agency as such, and yet paradoxically enhances the capacities of the embodied self in the mode of assemblage.

10. In much of my published work with Janet Price, we have adopted a DeleuzoGuattarian undecidable voice, adapting the opening lines to *A Thousand Plateaus* (1987): *The two of us wrote together. Since each of us is several, there*

was already quite a crowd. The purpose has been both a theoretical commitment to thinking through the intercorporeality of authorship, which starts
long before publication (see Potts and Price 1995), and the more pragmatic
desire to overcome the restriction that the disability experienced by one of
us – which not only often causes exhaustion, but also inconsistently affects
speech patterns – has meant that though the process as a whole has been
collaborative, that may not be true of all its parts. In consequence, my practice in 'quoting' Janet Price should not be taken in a conventional sense of
reproducing exact words written or spoken by her.

11.   I would reiterate also what Sara Ahmed has to say about reworking texts from
outside one's own specialist area: 'The promise of interdisciplinary scholarship is that the failure to return texts to their histories will do something'
(2006: 22).

12.   As before, in referring to ethics, I mean not a set of moral principles or rules
of conduct at the disposal of the sovereign individual, but the way in which
the relation between self and other is negotiated. Ethics on this reading is
always a matter of responsibility to and for the other, both in her presence
and absence.

13.   Affective labour – the work of the production and manipulation of affect – is
the term Hardt and Negri (2000: 292–3) give to the third of the three forms
of *immaterial* labour. It relates to work that influences the creation of feelings
of ease and comfort, such as nursing and social care, as well as areas such as
entertainment and other personal service work. Whilst its primary output is
affective, it encompasses labour that is both deeply entwined with the body
and productive of the corporeal.

14.   See Toynbee (2003) for a fuller account of a partially hidden care and service
economy that exploits women in particular.

15.   Neither the major trope of schizophrenia, nor other minor moments that
circulate around various signifiers of disability, could be said to engage with
the disabled body as such.

16.   I cannot agree with Dianne Currier's insistence on a distinction between
prosthetic hybridity and assemblages (Currier 2003). While her claim that
prostheses only facilitate completion within a logic of unity may be justified
with reference to *intent*, my understanding is that, regardless of the objec-
tives, all such conjunctions effect – at very least – phenomenological change
to the point of producing radically new assemblages.

17.   See Boron (2005).

Bibliography

Agamben, Giorgio (1998) *Homo Sacer: Sovereign Power and Bare Life*. Trans. D. Heller-Roazen. Stanford: Stanford University Press.

Aguilera, Raymond (2000) 'Disability and delight: Staring back at the devotee community', *Sexuality and Disability* 18: 255–61.

Ahmed, Sara (2006) *Queer Phenomenology: Orientations, Objects, Others*. Durham: Duke University Press.

Aldrovandi, Ulisse (1642) *Monstrorum Historia*. Bologna: Tebaldini.

Anzieu, Didier (1989) *The Skin Ego*. New Haven: Yale University Press.

Armer, William (2005). *In the Shadow of Genetics: An Analysis of Eugenic Influences on Twentieth Century Social Policy for Disabled People in European and North American Societies*. Unpublished PhD thesis, University of Leeds, Yorkshire, United Kingdom.

Augustine (1972) *City of God*. Trans. H. Betterson. Harmondsworth: Penguin Books.

Bacchi, Carol and Beasley, Chris (2002) 'Citizen bodies: Is embodied citizenship a contradiction in terms?', *Critical Social Policy* 22: 324–52.

Barron, Anne (2000) 'Spectacular jurisprudence', *Oxford Journal of Legal Studies* 20: 301–15.

Bauman, Zygmunt (1989) *Modernity and the Holocaust*. Cambridge: Polity.

—— (1998) *Globalization: The Human Consequences*. Cambridge: Polity Press.

Beail, Nigel and Warden, Sharon (1995) 'Sexual abuse of adults with learning disabilities', *Journal of Intellectual Disability Research* 39: 382–7.

Begum, Nasa (1992) 'Disabled women and the feminist agenda' in H. Hinds, A. Phoenix and J. Stacey (eds) *Working Out: New Directions for Women's Studies*. Brighton: Falmer Press.

Beresford, Peter and Holden, Chris (2000) 'We have choices: Globalization and welfare user movements', *Disability & Society* 15. 7: 973–89.

Bergoffen, Deborah (2000) 'Simone de Beauvoir: Disrupting the metonomy of gender' in D. Olkowski (ed.) *Resistance, Flight, Creation: Feminist Enactments of French Philosophy*. Ithaca, NY: Cornell University Press.

Berlant, Lauren (1997) *The Queen of America Goes to Washington City: Essays on Sex and Citizenship*. Durham, NC: Duke University Press.

Berlant, Lauren and Warner, Michael (1998) 'Sex in public', *Critical Inquiry* 24. 2: 547–66.

Bérubé, Michael (2002) 'Forward: Side shows and back bends' in L. Davis (ed.) *Bending over Backwards: Disability, Dismodernism and Other Difficult Positions*. New York: New York University Press.

Block, Pamela (2000) 'Sexuality, fertility and danger: Towards twentieth century images of women with cognitive disabilities', *Sexuality & Disability* 18. 4: 239–54.

Boaistuau, Pierre (1560) *Histoires prodigieuses les plus mémorables*. Paris: Vincent Sertenas.

193

Boothby, Richard (1991) *Death and Desire: Psychoanalytic Theory in Lacan's Return to Freud*. London: Routledge.
Boron, Atilio (2005) *'Empire' and Imperialism: A Critical Reading of Michael Hardt and Antonio Negri*. Oxford: Zed Books.
Boylan, Esther (1991) *Women and Disability*. London: Zed Books.
Brady, Susan, Britton, John and Grover, Sonia (2001) 'The sterilization of girls and young women in Australia'. Retrieved October 3, 2006, from the Human Rights and Equal Opportunity Commission website: http://www.hreoc.gov.au/disability_rights/sterilisation/intro.html.
Braidotti, Rosi (1991) *Patterns of Dissonance*. Cambridge: Polity Press.
—— (1994) *Nomadic Subjects*. New York: Columbia University Press.
—— (2005–2006) 'Affirming the affirmative: On nomadic affectivity', *Rhizomes* 11–12: np. Retrieved February 3, 2008, from www.rhizomes.net/issue11/braidotti/index.html.
—— (2006) *Transpositions: On Nomadic Ethics*. Cambridge: Polity Press.
Brown, Wendy (2002) 'Suffering the paradoxes of rights' in Wendy Brown and Janet Halley (eds) *Left Legalism/Left Critique*. Durham: Duke University Press.
Brown, Wendy and Halley, Janet (eds) (2002) *Left Legalism/Left Critique*. Durham: Duke University Press.
Brown, Jan and Russell, Sarah (2005) 'My home, your workplace: People with physical disability negotiate their sexual health without crossing professional boundaries', *Disability & Society* 20: 375–88.
Buffon, C. L. L. de (1749) *Sur les Monstres*. Paris.
Butler, Judith (1989) 'Sexual ideology and phenomenological description: A feminist critique of Merleau-Ponty's *Phenomenology of Perception*' in J. Allen and I. M. Young (eds) *The Thinking Muse: Feminism and Modern French Philosophy*. Bloomington: Indiana University Press.
—— (1990) *Gender Trouble: Feminism and the Subversion of Identity*. London: Routledge.
—— (1993) *Bodies that Matter: On the Discursive Limits of 'Sex'*. London: Routledge.
—— (2002) 'Is kinship always already heterosexual?' in W. Brown and J. Halley (eds) *Left Legalism/Left Critique*. Durham, NC: Duke University Press.
Caputo, John (2000) *More Radical Hermeneutics: On Not Knowing Who We Are*. Bloomington: Indiana University Press.
Carlson, Licia (2001) 'Cognitive ableism and disability studies: Feminist reflections on the history of mental retardation', *Hypatia* 16. 4: 124–46.
Cartwright, Lisa and Goldfarb, Brian (2006) 'On the subject of neural and sensory prostheses' in M. Smith and J. Morra (eds) *The Prosthetic Impulse from a Posthuman Present to a Biocultural Future*. Cambridge, MA: MIT Press.
Clare, Eli (2001) 'Stolen bodies, reclaimed bodies: disability and queerness', *Public Culture* 13. 3: 359–66.
Corker, Mairian (2001) 'Sensing disability', *Hypatia* 16. 4: 34–52.
Corker, Mairian and Shakespeare, Tom (eds) (2002) *Disability/Postmodernism: Embodying Disability Theory*. London: Continuum.
Cornell, Drucilla (1991) *Beyond Accommodation: Ethical Feminism, Deconstruction and the Law*. London: Routledge.
—— (1995) *The Imaginary Domain*. London: Routledge.

Couser, Thomas (1997) *Recovering Bodies: Illness, Disability and Life Writing*. Wisconsin: University of Wisconsin Press.

Currier, Dianne (2003) 'Feminist technological futures: Deleuze and body/technology assemblages', *Feminist Theory* 4. 3: 321–38.

Curry, Mary Ann, Hassouneh-Phillips, Dena and Johnston-Silverberg, Anne (2001) 'Abuse of women with disabilities: An ecological model and review', *Violence Against Women* 7: 60–79.

Davis, Lennard (1997) 'Nude Venuses, Medusa's body, and phantom limbs' in D. Mitchell and S. Snyder (eds) *The Body and Physical Difference*. Ann Arbor, MI: University of Michigan Press.

—— (2002) *Bending over Backwards: Disability, Dismodernism and Other Difficult Positions*. New York: New York University Press.

Dean, Tim (2000) *Beyond Sexuality*. Chicago: University of Chicago Press.

Deleuze, Gilles (1990) *The Logic of Sense*. Trans. M. Lester. London: Athlone Press.

—— (1994) 'Désir et plaisir', *Magazine Littéraire* 325 (October): 59–65.

—— (1995) *Negotiations 1972–1990*. Trans. M. Joughin. New York: Columbia University Press.

Deleuze, Gilles and Guattari, Félix (1984) *Anti-Oedipus: Capitalism and Schizophrenia*. Trans. R. Hurley. Minneapolis: Minnesota University Press.

—— (1987) *A Thousand Plateaus: Capitalism and Schizophrenia*. Trans. B. Massumi. Minneapolis: Minnesota University Press.

Deleuze, Gilles and Parnet, Claire (1987) *Dialogues*. Trans. H. Tomlinson and B. Habberjam. London: Athlone.

Derrida, Jacques (1988) 'Afterword: Toward an ethic of discussion' in *Limited Inc*. Trans. S. Weber, ed. G. Graff. Evanston, ILL: Northwestern University Press.

—— (1990) 'Force of law: The "Mystical Foundation of Authority"', Trans. M. Quaintance, *Cardozo Law Review* 11. 5–6: 919–1045.

—— (1995) *Points: Interviews, 1974–1994*. Trans. P. Kamuf *et al*, ed. E. Weber. Stanford, CA: Stanford University Press.

—— (1998) *Resistances of Psychoanalysis*. Trans. P. Kamuf. Stanford, CA: Stanford University Press.

—— (2000) 'Foreigner question' in *Of Hospitality: Anne Dufourmantelle Invites Jacques Derrida to Respond*. Trans. R. Bowlby. Stanford, CA: Stanford University Press.

Diedrich, Lisa (2005) 'Genealogies of disability: Historical emergences and everyday enactments', *Cultural Studies* 19. 6: 649–66.

—— (ed.) (2005) 'Genealogies of disability', *Cultural Studies* 19. 6, Special Issue.

Diprose, Rosalyn (1998) 'Sexuality and the clinical encounter' in M. Shildrick and J. Price (eds) *Vital Signs: Feminist Reconfigurations of the Bio/logical Body*. Edinburgh: Edinburgh University Press.

—— (2002) *Corporeal Generosity: On Giving with Nietzsche, Merleau-Ponty and Levinas*. New York: SUNY Press.

Disabled Peoples International (2000) 'Disabled people speak on the new genetics'. Retrieved September 29, 2006, from http://www.mindfully.org/GE/Disabled-People-Speak.htm.

Doel, Marcus (1999) *Poststructuralist Geographies: The Diabolical Art of Spatial Science*. Edinburgh: Edinburgh University Press.

Dongen, Els van and Elema, Riekje (2001) 'The art of touching: The culture of "body work" in nursing', *Anthropology and Medicine* 8. 2–3: 149–62.

Dowbiggin, Ian (2002) 'A rational coalition: Euthanasia, eugenics, and birth control in America, 1940–1970', *Journal of Policy History* 14. 3: 223–60.

Earle, Sarah (1999) 'Facilitated sex and the concept of sexual need: Disabled students and their personal assistants', *Disability & Society* 14: 309–23.

Ehrenreich, Barabara and Hochschild, Arlie Russell (eds) (2003) *Global Women: Nannies, Maids and Sex Workers in the New Economy.* London: Granta.

Erickson, Loree (2007) 'Revealing femmegimp: A sex positive reflection on sites of shame as sites of resistance for people with disabilities', *Atlantis: A Women's Studies Journal* 31. 2: np.

Fenton, Edward (1569) *Certaine Secrete Wonders of Nature.* London: Henry Bynneman.

Ferrell, Robyn (1996) *Passion in Theory: Conceptions of Freud and Lacan.* London: Routledge.

Fiedler, Leslie (1978) *Freaks: Myths and Images of the Secret Self.* New York: Simon and Schuster.

Foucault, Michel (1965) *Madness and Civilization: A History of Insanity in the Age of Reason.* Trans. R. Howard. New York: Random House.

——— (1973) *The Birth of the Clinic: An Archaeology of Medical Perception.* Trans. A. Sheridan. London: Tavistock.

——— (1977a) 'Nietzsche, Genealogy, History' in D. Bouchard (ed.) *Michel Foucault: Language, Counter-memory, Practice: Selected Essays and Interviews.* Trans. A. M. Sheridan Smith. Ithaca, NY: Cornell University Press.

——— (1977b) 'Intellectuals and power' in D. Bouchard (ed.) *Michel Foucault: Language, Counter-Memory, Practice.* Ithaca, NY: Cornell University Press.

——— (1977c) *Discipline and Punish: The Birth of the Prison.* Trans. A. Sheridan. London: Allen Lane.

——— (1979a) *History of Sexuality, Vol.1.* Trans. R. Hurley. London: Allen Lane.

——— (1979b) 'Governmentality'. *Ideology and Consciousness* 6: 5–29.

——— (1980) *Power/Knowledge. Selected Interviews and Other Writings 1972–1977.* ed. C. Gordon. Brighton: Harvester Press.

——— (1988a) 'The ethic of care for the self as a practice of freedom' in J. Bernauer and D. Rasmussen (eds) *The Final Foucault.* Cambridge, MA: MIT Press.

——— (1988b) *Politics, Philosophy, Culture: Interview and other writings 1977–1984.* Trans. A. Sheridan, ed. L. Kritzman. London: Routledge.

——— (1997) 'Sexual choice, sexual act' in P. Rabinow (ed.) *Ethics, Subjectivity and Truth: The Essential Works of Michel Foucault.* Trans. R. Hurley. New York: New Press.

——— (2003a) *Abnormal.* Trans. G. Burchell. New York: Picador.

——— (2003b) *Society Must Be Defended.* Trans. D. Macey, ed. M. Bertani and A. Fontana. New York: Picador.

Freud, Sigmund [1905] (1962) *Three Essays on the Theory of Sexuality.* Trans. J. Strachey. London: Basic Books.

——— [1919] (2003) *The Uncanny.* Trans. D. McLintock. London: Penguin Books.

Friedlander, Henry (1995) *The Origins of Nazi Genocide: From Euthanasia to the Final Solution.* Chapel Hill: University of North Carolina Press.

Friedman, John Block (1981) *The Monstrous Races in Medieval Art and Thought.* Cambridge, MA: Harvard University Press.

Fritsch, Kelly (2008) 'Nothing is immobile: Disability, relationality and unhinging possibilities'. Unpublished MA dissertation for Critical Disability Studies, York University, Toronto.

Furner, Ben (2006, December 6) Introduction. *The Guardian* (Independent living: Making equality a reality, supplement to *SocietyGuardian* section) 1–8.

Gallagher, H. G. (1990) *By Trust Betrayed: Patients, Physicians and the License to Kill in the Third Reich*. Clearwater, FL: Vandamere Press.

Garland-Thomson, Rosemarie (ed.) (1996) *Freakery: Cultural Spectacles of the Extraordinary Body*. New York: New York University Press.

—— (2002) 'Integrating disability, transforming feminist theory', *NWSA Journal* 14. 3: 1–32.

Gatens, Moira (1996) *Imaginary Bodies: Ethics, Power, and Corporeality*. London: Routledge.

Gerodetti, Natalia (2006) 'From science to social technology: Eugenics and politics in twentieth century Switzerland', *Social Politics: International Studies in Gender, State and Society* 13: 59–88.

Ghai, Anita (2003) *(Dis)embodied Forms: Issues of Disabled Women*. New Delhi: Shakti Books.

Gibson, Barbara E. (2006) 'Disability, connectivity and transgressing the autonomous body', *Journal of Medical Humanities* 27: 187–96.

Giddens, Antony (1990) *The Consequences of Modernity*. Cambridge: Polity Press.

Glover, Jonathan (2000) *Humanity: A Moral History of the Twentieth Century*. New Haven: Yale University Press.

Goering, Sara (2002) 'Beyond the medical model? Disability, formal justice and the exception for the "profoundly impaired"', *Kennedy Institute of Ethics Journal* 12. 4: 373–88.

Goodley, Dan and Roets, Griet (2008) 'The (be)comings and goings of "developmental disabilities": The cultural politics of impairment', *Discourse: Studies in the Cultural Politics of Education* 29. 2: 239–55.

Goodrich, Peter (1987) *Legal Discourse: Studies in Linguistics, Rhetoric and Legal Analysis*. New York: St. Martins Press.

—— (1992) 'Critical legal studies in England: Prospective histories', *Oxford Journal of Legal Studies* 12: 195–236.

—— (1995) *Oedipus Lex: Psychoanalysis, History, Law*, Berkeley, CA: University of California Press.

Gould, George M. and Pyke, Walter L. (1897) *Anomalies and Curiosities of Medicine*. London: Rebman Publishing.

Grosz, Elizabeth (1994a) *Volatile Bodies: Towards a Corporeal Feminism*. Bloomington: Indiana University Press.

—— (1994b) 'A thousand tiny sexes: Feminism and rhizomatics' in C. Boundas and D. Olkowski (eds) *Gilles Deleuze and the Theatre of Philosophy*. London: Routledge.

—— (1995) *Space, Time, and Perversion: Essays on the Politics of Bodies*. London: Routledge.

—— (2006) 'Naked' in M. Smith and J. Morra (eds) *The Prosthetic Impulse from a Posthuman Present to a Biocultural Future*. Cambridge, MA: MIT Press.

Guattari, Félix (1996) *Soft Subversions*. ed. S. Lotringer. New York: Semiotext(e).

Halperin, David (1996) 'The queer politics of Michel Foucault' in D. Morton (ed.) *The Material Queer*. Boulder: Westview Press.

Hannabach, Cathy (2007) 'Anxious embodiment, disability, and sexuality: A response to Margrit Shildrick', *Studies in Gender and Sexuality* 8. 3: 253–61.

Harari, Roberto (2001) *Lacan's Seminar on 'Anxiety'*. New York: Other Press.

Haraway, Donna (1991) 'A cyborg manifesto: Science, technology, and socialist-feminism in the late twentieth century' in *Simians, Cyborgs, and Women: The Reinvention of Nature*. London: Free Association Books.

——— (1992) 'The promises of monsters: A regenerative politics for inappropriate/d others' in Lawrence Grossberg *et al* (eds) *Cultural Studies*. London: Routledge.

——— (1994) 'A game of cat's cradle: Science studies, feminist theory, cultural studies', *Configurations* 2. 1: 59–71.

Hardt, Michael and Negri, Antonio (2000) *Empire*. Cambridge, MA: Harvard University Press.

Harvey, David (1989) *The Condition of Postmodernity*. Oxford: Basil Blackwell.

Hershey, Laura (2000) 'Women with disabilities: Health, reproduction, and sexuality' in *International Encyclopedia of Women: Global Women's Issues and Knowledge*. New York: Routledge.

Hewson, Barbara (2001) 'Killing off Mary: Was the Court of Appeal right?', *Medical Law Review* 9. 3: 281–98.

Hoche, Alfred and Binding, Karl (1920) *Die freigabe der vernichtung lebensunwertem lebens*. Leipzig, Germany: K. Felix Meiner Verlag. ['Permitting the destruction of unworthy life', Trans. W. E. Wright. *Issues in Law and Medicine* 1992, 8. 2: 231–65.]

Hoogvelt, Ankie (1997) *Globalisation and the Postcolonial World*. London: Macmillan Press.

Hughes, Bill, McKie, Linda, Hopkins, Debra and Watson, Nick (2005) 'Love's labours lost? Feminism, the disabled people's movement and an ethic of care', *Sociology* 39. 2: 259–75.

Hurst, Rachel (2003) 'Conclusion: Enabling or disabling globalization' in J. Swain, S. French and C. Cameron (eds) *Controversial Issues in a Disabling Society*. Buckingham: Open University Press.

Iwakuma, Miho (2002) 'The body as embodiment: An investigation of the body by Merleau-Ponty' in M. Corker and T. Shakespeare (eds) *Disability/ Postmodernism: Embodying Disability Theory*. London: Continuum.

Irigaray, Luce (1985) *Speculum of the Other Woman*. Trans. G. C. Gill. Ithaca, NY: Cornell University Press.

——— (1993) 'The intertwining – The chiasm' in *An Ethics of Sexual Difference*. Trans. C. Burke and G. C. Gill. Ithaca, NY: Cornell University Press.

Jacob, François (1989) *The Logic of Life: A History of Heredity*. Trans. B. E. Spillmann. Harmondsworth: Penguin Books.

Job, Jennifer (2004) 'Factors involved in the ineffective dissemination of sexuality information to individuals who are deaf or hard of hearing', *American Annals of the Deaf* 149: 264–73.

Kafer, Alison (2003) 'Compulsory bodies: Reflections on Heterosexuality and able-bodiedness', *Journal of Women's History* 15. 3: 77–89.

——— (2004) 'Inseparable: Gender and disability in the amputee-devotee community' in B. Smith and B. Hutchison (eds) *Gendering Disability*. New Brunswick, NJ: Rutgers University Press.

Karpin, Isabel (2005) 'Genetics and the legal conception of self' in M. Shildrick and R. Mykitiuk (eds) *Ethics of the Body: Postconventional Challenges*. Cambridge, MA: MIT Press.

Kennedy, Margaret (1996) 'Sexual abuse and disabled children' in J. Morris (ed.) *Encounters with Strangers: Feminism and Disability*. London: The Women's Press.

Keywood, Kirsty (2001) '"I'd rather keep him chaste." Retelling the story of steril- isation, learning disability and (non)sexed embodiment', *Feminist Legal Studies* 9: 185–94.

Klein, Naomi (2002) *Fences and Windows: Dispatches from the Front Lines of the Globalization Debate*. London: Flamingo.

Kristeva, Julia (1982) *Powers of Horror: An Essay on Abjection*. New York: Columbia University Press.

——— (1990) *Strangers to Ourselves*. London: Harvester Wheatsheaf.

——— (1996) *Julia Kristeva Interviews*. ed. R. Mitchell Guberman. New York: Columbia University Press.

——— (2000) *The Sense and Non-sense of Revolt: The Power and Limits of Psycho- analysis*. Trans. J. Herman. New York: Columbia University Press.

Lacan, Jacques (1977a) 'Aggressivity in psychoanalysis' in *Écrits: A Selection*. Trans. A. Sheridan. New York: W. W. Norton.

——— (1977b) 'The mirror stage as formative of the function of the I' in *Écrits: A Selection*. Trans. A. Sheridan. New York: W. W. Norton.

——— (1981) *The Four Fundamental Concepts of Psychoanalysis*. Trans. A. Sheridan. New York: W. W. Norton.

——— (1988) *The Seminar of Jacques Lacan*. Book II, Trans. S. Tomaselli, ed. J.-A. Miller. New York: W. W. Norton.

Lingis, Alphonso (1985) *Libido: The French Existential Theories*. Bloomington: Indiana University Press.

Longmore, Paul K. (1997) 'Conspicuous contribution and American cultural dilemmas: Telethon rituals of cleansing and renewal' in D. Mitchell and S. Snyder (eds) *The Body and Physical Disability: Discourses of Disability*. Ann Arbor, MI: University of Michigan Press.

Lorraine, Tamsin (2000) 'Becoming-imperceptible as a mode of self-presentation: A feminist model drawn from a Deleuzian line of flight' in D. Olkowski (ed.) *Resistance, Flight, Creation: Feminist Enactments of French Philosophy*. Ithaca, NY: Cornell University Press.

Lupton, Deborah (1999) *Risk*. London: Routledge.

MacKinnon, Catharine (1987) *Feminism Unmodified*. Cambridge, MA: Harvard University Press.

Malone, Kareen and Kelly, Shannon (2004) 'The transfer from the clinical and the social and back', *Psychoanalysis, Culture & Society*. 9. 1: 23–32.

Marks, Deborah (1999) *Disability: Controversial Debates and Psychosocial Perspec- tives*. London: Routledge.

Mayhew, Les (2003) 'Disability – global trends and international per- spectives', *Innovation: The European Journal of Social Science Research* 16. 1: 3–28.

McRuer, Robert (2003) 'As good as it gets: Queer theory and critical disability', *GLQ: A Journal of Lesbian and Gay Studies* 9. 1–2: 79–105.

——— (2006) *Crip Theory: Cultural Signs of Queerness and Disability*. New York: New York University Press.

McRuer, Robert and Wilkerson, Abby (2003) 'Cripping the (queer) nation', *GLQ A Journal of Lesbian and Gay Studies* 9. 1–2: 1–23.

Meekosha, Helen (1998) 'Body battles: Bodies, gender and disability' in T. Shakespeare (ed.) *The Disability Reader*. London: Cassell.

——— (2004) 'Drifting down the gulf stream: Navigating the cultures of disability studies', *Disability & Society* 19. 7: 721–33.

Menninghaus, Winfried (2003) *Disgust: Theory and History of a Strong Sensation*. Trans. H. Eiland and J. Golb. New York: SUNY Press.

Merleau-Ponty, Maurice (1962) *Phenomenology of Perception*. Trans. C. Smith. London: Routledge & Kegan Paul.

——— (1964) *The Primacy of Perception*. Evanston: Northwestern University Press.

——— (1968) *The Visible and the Invisible*. Evanston: Northwestern University Press.

Michalko, Rod (1999) *The Two in One: Walking with Smokie, Walking with Blindness*. Philadelphia: Temple University Press.

Michalowski, Sabine (2001) 'Reversal of fortune – *Re A (Conjoined Twins)* and beyond: Who should make treatment decisions on behalf of young children?', *Health Law Journal* 9: 149–69.

Minsky, Rosalind (ed.) (1996) *Psychoanalysis and Gender*. London: Routledge.

Mitchell, David and Snyder, Sharon (eds) (1997) *The Body and Physical Disability: Discourses of Disability*. Ann Arbor, MI: University of Michigan Press.

——— (2000) *Narrative Prosthesis: Disability and the Dependencies of Discourse*. Ann Arbor, MI: University of Michigan Press.

——— (2003) 'The eugenic Atlantic: Race, disability and the making of an international eugenic science, 1800–1945', *Disability & Society* 18. 7: 843–64.

Montagu, Ashley (1971) *Touching: The Human Significance of the Skin*. New York: Columbia University Press.

Morris, Jenny (ed.) (1996) *Encounters with Strangers: Feminism and Disability*. London: The Women's Press.

Mostert, Mark (2002) 'Useless eaters: Disability as genocidal marker in Nazi Germany', *Journal of Special Education* 36: 157–70.

Munro, Vanessa (2001) 'Square pegs in round holes: The dilemma of conjoined twins and individual rights', *Social and Legal Studies* 10. 4: 459–82.

Nedelsky, Jennifer (1989) 'Reconceiving autonomy – sources, thoughts and possibilities', *Yale Journal of Law and Feminism* 1: 7–36.

Nietzsche, Friedrich (1982) *Daybreak*. Cambridge: Cambridge University Press.

Nosek, Margaret, A., Howland, Carol, Rintala, Diana, Young, Mary Ellen and Chanpong, Gail (2001) 'National study of women with physical disabilities: Final report', *Sexuality and Disability* 19. 1: 5–39.

Notes from Nowhere Collective (2003) *We are Everywhere: The Irresistible Rise of Global Anti-capitalism*. London: Verso.

Oliver, Michael (1990) *The Politics of Disablement*. Basingstoke: Macmillan.

——— (1996) *Understanding Disability: From Theory to Practice*. Basingstoke: Macmillan.

Olkowski, Dorothea (2002) 'Flesh to desire: Merleau-Ponty, Bergson, Deleuze', *Strategies* 15. 1: 11–24.

Ott, Katherine (2002) 'The sum of its parts' in K. Ott, D. Serlin and S. Mihm (eds) *Artificial Parts, Practical Lives: Modern Histories of Prosthetics*. New York: New York University Press.

Ott, Katherine, Serlin, David and Mihn, Stephen (eds) (2002) *Artificial Parts, Practical Lives*. New York: New York Univeristy Press.

Overboe, James (1999) '"Difference in itself": Validating disabled people's lived experience', *Body & Society* 5. 4: 17–29.

—— (2007a) 'Disability and genetics: Affirming the bare life the state of exception', *Canadian Review of Sociology and Anthropology* 44: 219–35.

—— (2007b) 'Vitalism: Subjectivity exceeding racism, sexism and (psychiatric) ableism', *Wagadu* 4: 23–34.

Panzarino, Connie (1996) 'To my other bodies' in S. Tremain (ed.) *Pushing the Limits: Disabled Dykes Produce Culture*. Toronto: Women's Press, 85–6.

Paré, Ambroise (1982) *On Monsters and Marvels*. Trans. J. L. Pallister. Chicago: Chicago University Press.

Parisi, Luciana (2004) *Abstract Sex: Philosophy, Bio-technology and the Mutations of Desire*. London: Continuum.

Park, Katharine and Daston, Lorraine (1981) 'Unnatural conceptions: The study of monsters in sixteenth and seventeenth century France and England', *Past and Present* 92: 20–54.

Parritt, Simon (2005) 'Sex: The final frontier', *Disability Now*. Retrieved October 3, 2006, from http://www.disabilitynow.org.uk/timetotalksex/survey_006.htm.

Paterson, Jennifer (2007) 'Beyond clinical and heteronormative encounters between disability and sexuality: Kinky agency and resistance'. Unpublished MA dissertation in Critical Disability Studies, York University, Toronto.

Paterson, Kevin and Hughes, Bill (1999) 'Disability studies and phenomenology: The carnal politics of everyday life', *Disability & Society* 14: 597–610.

Paterson, Mark (2005) 'Affecting touch: Towards a "felt" phenomenology of therapeutic touch' in J. Davidson, L. Bondi and M. Smith (eds) *Emotional Geographies*. Aldershot: Ashgate.

Pender, Stephen (1996) '"No monsters at the resurrection": Inside some conjoined twins' in J. J. Cohen (ed.) *Monster Theory*. Minneapolis: Minnesota University Press.

Pfeiffer, Davi (1994) 'Eugenics and disability discrimination', *Disability & Society* 9: 481–99.

—— (2002) 'A comment on the social model(s)', *Disability Studies Quarterly* 22. 4: 234–5.

Phelan, Shane (2001) *Sexual Strangers. Gays, Lesbians, and Dilemmas of Citizenship*. Philadelphia: Temple University Press.

Potts, Tracey and Price, Janet (1995) 'Out of the blood and spirit of our lives: The place of the body in academic feminism' in L. Morley and V. Walsh (eds) *Feminist Academics: Creative Agents for Change*. London: Taylor and Francis.

Prendergast, Catherine (2008) 'The unexceptional schizophrenic: A post-postmodern introduction', *Journal of Literary Disability* 2. 1: 55–62.

Price, Janet and Shildrick, Margrit (1998) 'Uncertain thoughts on the dis/abled body' in M. Shildrick and J. Price (eds) *Vital Signs*. Edinburgh: Edinburgh University Press.

Price, Janet and Shildrick, Margrit (2001) 'Bodies together: Touch, ethics, and disability' in M. Corker and T. Shakespeare (eds) *Disability/Postmodernism: Embodying Disability Theory*. London: Continuum.

Ratzka, Adolf (2002) 'Sexuality and people with disabilities: What experts are not aware of'. *Institute of Independent Living*. Retrieved February 12, 2008, from http://www.independentliving.org/docs5/Sexuality.html.

Razack, Sherene (1994) 'From consent to responsibility, from pity to respect: Subtexts in cases of *sexual* violence involving girls or women with developmental disabilities', *Law and Social Inquiry* 19: 891–922.

Richardson, Diane (2000) 'Constructing sexual citizenship: Theorizing sexual rights', *Critical Social Policy* 20: 105–35.

––––––– (2004) 'Locating sexualities: From here to normality', *Sexualities* 7: 391–411.

Rioux, Marcia and Carbert, Anne (2004) 'Human rights and disability: The international context, *Journal on Developmental Disabilities* 10. 2: 1–13.

Rohrer, Judy (2005) 'Towards a full-inclusion feminism: A feminist deployment of disability analysis', *Signs* 31.1: 34–63.

Rose, Jacqueline (1993) *Why War? Psychoanalysis, Politics and the Return to Melanie Klein*. Oxford: Blackwell.

Salih, Sara and Butler, Judith (eds) (2004) *The Judith Butler Reader*. Oxford: Blackwell.

Sedgwick, Eve Kosofsky (1994) *Tendencies*. London: Routledge.

Seidman, Steven (2001) 'From identity to queer politics: Shifts in normative heterosexuality and the meaning of citizenship', *Citizenship Studies* 5: 321–28.

––––––– (2002) *Beyond the Closet: The Transformation of Gay and Lesbian Life*. London: Routledge.

Serlin, David (2006) 'Disability, masculinity, and the prosthetics of war, 1945–2005' in M. Smith and J. Morra (eds) *The Prosthetic Impulse from a Posthuman Present to a Biocultural Future*. Cambridge, MA: MIT Press.

Seymour, M. C. (ed.) (1967) *Mandeville's Travels*. Oxford: Clarendon Press.

Shakespeare, Tom (1999) 'The sexual politics of disabled masculinity', *Sexuality and Disability* 17. 1: 53–64.

––––––– (2000) 'Disabled sexuality: Toward rights and recognition', *Sexuality and Disability* 18. 3: 159–66.

––––––– (2005) 'Sex, death and stereotypes: Disability in *Sick* and *Crash*' in G. Harper and A. Moor (eds) *Signs of Life: Medicine and Cinema*. London: Wallflower Press.

Shakespeare, Tom, Gillespie-Sells, Kath and Davies, Dominic (1996) *The Sexual Politics of Disability: Untold Desires*. London: Cassell.

Sheldon, Sally and Wilkinson, Steve (1997) 'Conjoined twins – the ethics and legality of sacrifice', *Medical Law Review* 5. 2: 149–71.

Sherry, Mark (2004) 'Overlaps and contradictions between queer theory and disability studies', *Disability & Society* 19: 769–83.

Shildrick, Margrit (1997) *Leaky Bodies and Boundaries: Feminism, Postmodernism and (Bio)ethics*. London: Routledge.

––––––– (2002) *Embodying the Monster: Encounters with the Vulnerable Self*. London: Sage.

––––––– (2004a) 'Silencing sexuality: The regulation of the disabled body' in J. Carabine (ed.) *Sexualities*. Buckingham and Bristol: Open University/Policy Press.

—— (2004b) 'Queering performativity: Disability after Deleuze', *SCAN: Journal of Media Arts Culture* 3. 1: np. Retrieved December 15, 2004 from http://www. scan.net.au/scan/journal/display.php?journal_id=36.

—— (2005a) 'The disabled body: Genealogy and undecidability', *Cultural Studies* 19: 755–70.

—— (2005b) 'Unreformed bodies: Normative anxiety and the denial of pleasure', *Women's Studies: An Interdisciplinary Journal* 34: 3–4: 327–44.

—— (2008) 'Deciding on death: Conventions and contestations in the context of disability', *Journal of Bioethical Inquiry* (Special issue: 'Disability and Bioethics') 5. 2–3: 209–19.

Shildrick, Margrit and Price, Janet (1996) 'Breaking the boundaries of the broken body: Mastery, materiality and ME', *Body and Society* 2. 4: 93–113.

—— (2001) 'Bodies together: Touch, ethics and disability' in M. Corker and T. Shakespeare (eds) *Disability/Postmodernism: Embodying Disability Theory*. London: Continuum.

—— (2005–2006) 'Deleuzian connections and queer corporealities: Shrinking global disability', *Rhizomes* 11–12: np. Retrieved February 3, 2008, from www. rhizomes.net/issue11/shildrickprice/index.html.

Shue, Karen and Flores, Ana (2002) 'Whose sex is it anyway? Freedom of exploration and expression of sexuality of an individual living with a brain injury in a supported independent living environment', *Disability Studies Quarterly* 22. 4: 59–72.

Shuttleworth, Russell (2002) 'Defusing the adverse context of disability and desirability as a practice of the self for men with cerebral palsy' in M. Corker and T. Shakespeare (eds) *Disability/Postmodernism: Embodying Disability Theory*. London: Continuum.

Siebers, Tobin (2008) *Disability Theory*. Ann Arbor, MI: University of Michigan Press.

Snyder, Sharon and Mitchell, David (2001) 'Re-engaging the body: Disability studies and the resistance to embodiment', *Public Culture* 13: 367–89.

Sobchack, Vivian (2006) 'A leg to stand on: Prosthetics, metaphor and materiality' in M. Smith and J. Morra (eds) *The Prosthetic Impulse From a Posthuman Present to a Biocultural Future*. Cambridge, MA: MIT Press.

Stiegler, Bernard (1998) *Technics and Time: The Fault of Epimetheus No.1*. Trans. R. Beardsworth and G. Collins. Stanford: University of Stanford Press.

Stienstra, Deborah (2002) 'Disabling globalisation: Rethinking global political economy with a disability lens', *Global Society* 16. 2: 109–21.

Stiker, Henri-Jacques (1999) *A History of Disability*. Trans. W. Sayers. Ann Arbor, MI: University of Michigan Press. (Originally published as *Corps infirmes et sociétiés*. 1982, Editions Aubier Montaigne.)

—— (2007) 'The contribution of human sciences to the field of disability in France over recent decades', *Scandinavian Journal of Disability Research* 9. 3–4: 146–59.

Sullivan, Nikki (2008) 'Dis-orienting paraphilias? Disability, desire, and the question of (bio)ethics', *Journal of Bioethical Inquiry* 5: 183–92.

Tepper, Mitch (1999) 'Letting go of restrictive notions of manhood: Male sexuality, disability and chronic Illness', *Sexuality and Disability* 17.1:37–52.

Thomas, Carol (1999) *Female Forms: Experiencing and Understanding Disability*. Buckingham: Open University Press.

Titchkosky, Tanya (2006) 'Policy, disability, reciprocity?' in M. A. McColl and L. Jongbloed (eds) *Disability and Social Policy in Canada*. Concord, ON: Captus Press.

Toynbee, Polly (2003) *Hard Work: Life in Low-Pay Britain*. London: Bloomsbury.

Tremain, Shelley (2000) 'Queering disabled sexuality studies', *Sexuality and Disability* 18. 4: 291–9.

——— (2001) 'On the government of disability', *Social Theory and Practice* 27: 617–36.

——— (2005) *Foucault and the Governmentality of Disability*. Ann Arbor, MI: University of Michigan Press.

Valios, Natalie (2001) 'Desire denied', *Community Care* 1391: 20–1.

Vansteenwegen A., Jans I. and Revell A. (2003) 'Sexual experience of women with a physical disability', *Sexuality and Disability* 21. 4: 283–90.

Warner, Michael (ed.) (1993) *Fear of a Queer Planet: Queer Politics and Social Theory*. Minneapolis: University of Minnesota Press.

——— (1999) *The Trouble with Normal: Sex, Politics, and the Ethics of Queer Life*. Cambridge, MA: Harvard University Press.

Watermeyer, Brian (2006) 'Disability and psychoanalysis' in B. Watermeyer, L. Swartz, T. Lorenzo, M. Schneider and M. Priestley (eds) *Disability and Social Change: A South African Agenda*. Cape Town: Human Sciences Research Council.

Watt, Helen (2001) 'Conjoined twins: Separation as mutilation', *Medical Law Review* 9. 3: 237–45.

Weeks, Jeffrey (1998) 'The sexual citizen', *Theory, Culture and Society* 15. 3–4: 35–52.

Weiss, Gail (1999) *Body Images: Embodiment as Intercorporeality*. London: Routledge.

Wendell, Susan (1996) *The Rejected Body: Feminist Philosophical Reflection on Disability*. London: Routledge.

——— (1997) 'Toward a feminist theory of disability' in L. Davis (ed.) *The Disabilities Studies Reader*. London: Routledge.

What YOU said in the survey (2005) *Disability Now*. Retrieved October 3, 2006, from http://www.disabilitynow.org.uk/timetotalksex/survey_007.htm.

White, Patrick (2003) 'Sex education: Or how the blind became heterosexual', *GLQ: A Journal of Lesbian and Gay Studies* 9: 133–47.

WHO (2004) 'Disability and Rehabilitation'. Retrieved October 3, 2006, from www.who.int/ncd/disability/index.htm.

Wiley, Juniper (1999) 'No Body is "Doing It": Cybersexuality' in J. Price and M. Shildrick (eds) *Feminist Theory and the Body*. Edinburgh: Edinburgh University Press.

Wilkinson, Abby (2002) 'Disability, sex radicalism, and political agency', *NWSA Journal* 14. 3: 33–57.

Williams, Linda (1999) 'The inside-out of masculinity: David Cronenberg's visceral pleasures' in M. Aron (ed.) *The Body's Perilous Pleasures: Dangerous Desires and Contemporary Culture*. Edinburgh: Edinburgh University Press.

Wilton, Robert (2003) 'Locating physical disability in Freudian and Lacanian psychoanalysis: Problems and prospects', *Social and Cultural Geography* 4. 3: 369–89.

World Bank (2004) website: http://webworldbank.org/WBSITE/EXTERNAL/TOPICS/EXTSOCIALPROTECTION/EXTDISABILITY/.

Young, Iris Marion (1990) *Throwing Like a Girl*. Bloomington: Indiana University Press.

Ziarek, Ewa Plonowska (1997) 'From euthanasia to the other of reason: Performativity and the deconstruction of sexual difference' in E. K. Feder, M. C. Rawlinson, and E. Zakin (eds) *Derrida and Feminism: Recasting the Question of Woman*. New York: Routledge.

Case cited

Re A (Conjoined Twins: Medical Treatment) 2001, 1 F.L.R. 1 (C.A.).

Index